Tendon-Sheath Transmitted Light and Slender Robots

Xingsong Wang · Qi Zhang

Tendon-Sheath Transmitted Light and Slender Robots

Tendon-Sheath Transmission Theory with Applications in Slender Robots, NOTES Robots, and Exoskeletons

东南大学出版社
SOUTHEAST UNIVERSITY PRESS

Springer

Xingsong Wang
School of Mechanical Engineering
Southeast University
Nanjing, China

Qi Zhang
School of Automation
Nanjing University of Information Science and Technology
Nanjing, China

ISBN 978-981-96-5580-9 ISBN 978-981-96-5581-6 (eBook)
https://doi.org/10.1007/978-981-96-5581-6

Jointly published with Southeast University Press
The print edition is not for sale in China (Mainland). Customers from China (Mainland) please order the print book from: Southeast University Press.
ISBN of the Co-Publisher's edition: 978-7-5766-2282-9

This Springer imprint is published by the registered company Springer Nature Singapore Pte Ltd.
The registered company address is: 152 Beach Road, #21-01/04 Gateway East, Singapore 189721, Singapore

Foreword

The advancement of robotics has continually pushed the boundaries of engineering innovation, calling for increasingly sophisticated mechanisms to enhance mobility, precision, and versatility. Among these innovations, tendon sheath transmission techniques have emerged as a powerful tool for addressing the challenges posed by designing compact, lightweight, and adaptable robotic systems.

This book presents a comprehensive exploration of tendon sheath transmission methods, delving into their principles, implementation strategies, and applications in diverse robotic platforms. From fundamental theoretical foundations to real-world case studies, it provides a unique blend of academic rigor and practical insights, bridging the gap between research and application.

The author has meticulously compiled his expertise to demonstrate how tendon sheath transmissions can enable advanced robotic functionalities, such as flexible actuation and remote power transmission, in domains ranging from medical devices to industrial automation. The inclusion of detailed examples and prototype robots further illustrates the versatility of these methods and their transformative potential in engineering and robotics.

I first met the author in the early 1990s when he was a Ph.D. student jointly under my supervision at Southeast University (SEU) in Nanjing, China. After completing his Ph.D., he joined the faculty at SEU, becoming my colleague. Together, we successfully secured numerous prestigious national research grants to address various challenges in robotics science and its applications. Although I later relocated to Hong Kong and New Zealand, our collaboration continued. In fact, one chapter of this book is a direct result of our joint efforts.

This book is a collection of authors' decades of diligent and smart works. I hope that this book will serve as an invaluable resource for researchers, practitioners, and students, inspiring new ideas and innovations in the rapidly evolving field of robotics. By offering both depth and clarity, this work highlights the promise of tendon sheath transmission techniques in shaping the next generation of robotic systems.

December 2024

Prof. Dr. Peter Xu
Chair in Mechatronics Engineering
The University of Auckland
Auckland, New Zealand

Preface

Slender robot is a kind of bionic and supernatural snake-like robot, which has a wide application prospects, such as searching and rescuing of life in natural disasters or large-scale accidents, inspecting and maintaining of narrow spaces inside large equipment, detecting of electrical or heating pipes inside building structures, as well as robot inspecting of human natural cavity and surgery. For decades, researchers have been exploring its bionic mechanism, innovating its structural design and control methods, which achieved numerous breakthroughs. However, till present, most existing slender robots still have bulky structures, and there is still a large gap between the motion performance and the growing application demands.

Tendon-sheath transmission, also known as bionic Tendon sheath transmission, is widely used in the brake and control transmission systems of bicycles, motorcycles and agricultural machines, with light weight, small size, and flexible structure. Since 2007, the first author of this book and his graduate students have been supported by four continuous National Natural Science Foundations in the field of tendon-sheath-driven robots, and achieved influential results in modeling of nonlinear transmission characteristics and precision motion control of tendon-sheath transmission, soft operation technology of tendon-sheath-driven slender robots, and design of tendon-sheath-driven mechanism to imitate artificial muscles, as well as tendon-sheath-driven soft exoskeleton and wearable rehabilitation devices. They published dozens of high-level journal papers in Science Robotics, IEEE Transactions on Industry Electronics, IEEE Transactions on Control Systems Technology, IEEE/ASME Transactions on Mechatronics, IEEE Transactions on Biomedical Engineering, etc., which forms a relatively systematic theory of design and control of tendon-sheath-driven mechanism, and its applications in slender robots and medical robots.

This book will systematically introduce the transmission theory of tendon-sheath-driven mechanism with structure innovations of slender robots of the author's group, which include structure designing of tendon-sheath-driven devices, modeling of force/displacement transfer characteristics of tendon-sheath-driven mechanism, precision motion control and error compensation of tendon-sheath-driven mechanisms, designing and modeling of tendon-sheath-driven slender robots using macro-micro-robot concept, morphological measurement and reconstruction theory of slender structures, designing and accurate motion control of micro-end actuators of slender robots, and tendon-sheath-driven lightweight exoskeleton robots.

The appearance of this book will fill the gap of lacking systematic theories for structure designing, characteristic modeling and motion control of this traditional tendon-sheath transmission mode. At the same time, it will apply modern material science, advanced mechanical structure design and manufacturing technology, advanced sensing and measurement technology, digital modeling and computer control technology to the design and control of tendon-sheath transmission mechanisms. It will provides theoretical and technical guidance for the application of tendon-sheath transmission technology to precision moving devices such as robots.

This book has a good reference value for scientific and technological personnel engaged in researching of robots, as well as for senior undergraduate and graduate students majoring in robotics and automation, mechatronic engineering, biomedical engineering, etc. It also can be a reference for engineers and technical personnel, as well as graduate students in the fields of special equipment maintenance and inspecting, disaster emergency management, intelligent construction and maintenance, medical equipment management, which will promote the integration and development of related technologies.

Nanjing, China
December 2024

Xingsong Wang

Acknowledgements The main content of this book is the result of the joint efforts of the first author and his graduate students, especially the excellent doctoral and master's students who have carried out their dissertations on topic of tendon-sheath transmission for more than 10 years. With the efforts of all of them, we have not only completed a variety of challenging innovative structural designs and engineering application attempts of slender robots, but also completed the theoretical achievements in tendon-sheath transmission theory. They brought their vigor and vitality of young people to the laboratory, added the courage and confidence of the first author's continuous innovation, and realized his dream of mutual growth through teaching and learning with his students. These students and their papers are listed below:

1. Chen Lin: "Modeling and Control of Tendon-Sheath Transmission and Its Application in Slender Soft Rescue and Inspection Robots," Ph.D. thesis, 2013
2. Wu Qingcong: "Key Technology Research on Tendon-Sheath Driven Gravity Balance Upper Limb Rehabilitation Exoskeleton," Ph.D. thesis, 2015
3. Xi Ruru: "Research on Tendon-Driven Semi-Passive Lower Limb Assistance Exoskeleton," Ph.D. thesis, 2016
4. Zhang Qi: "Design and Motion Control of Tendon-Driven Artificial Muscles for Exoskeleton Applications," Ph.D. thesis, 2020
5. Shen Donghua: "Design and Application Research on Biomimetic Snake Robots," Ph.D. thesis, 2021
6. Tian Fuxiang: "Research on Tendon-Sheath Driven Lightweight Slender Robot Technology," M.S. thesis, 2009
7. Wang Kai: "Design and Compliance Control of Tendon-Sheath Driven Slender Robot," M.S. thesis, 2014
8. Yuan Peng: "Development of a Tendon-Sheath Driven Proportional Mapping Control Robotic Arm," M.S. thesis, 2014
9. Zhang Guangxiao: "3D Reconstruction of Tendon-Sheath Driven Slender Robot Configurations," M.S. thesis, 2014
10. Shen Donghua: "Research on Expansion and Manipulation Techniques for Intra-body Surgical Robots," M.S. thesis, 2015
11. Li Jia: "Modeling and Compliance Control of Tendon-Sheath Transmission Systems," M.S. thesis, 2016
12. Chen Jianfei: "Research on Soft Rescue Robots," M.S. thesis, 2016
13. Gan Zhenbo: "Design of Lower Limb Rehabilitation Exoskeleton Mechanisms Driven by Tendon-Sheath Artificial Muscles," M.S. thesis, 2018
14. Qiu Junwei: "Research on a Compound Tendon-Sheath Artificial Muscle Lower Limb Single-Joint Rehabilitation Device," M.S. thesis, 2018
15. Shen Xiaopeng: "Design and Performance Study of Tendon-Sheath Driven Artificial Muscles," M.S. thesis, 2019
16. Li Haoming: "Development of a Tendon-Sheath Driven Six-Degree-of-Freedom Robotic Arm," M.S. thesis, 2021
17. Liu Lingyan: "Research on Multi-tendon-sheath Transmission Technology for Variable Stiffness Slender Robots," M.S. thesis, 2022

On the occasion of the publication of this book, I would like to express my sincere thanks to all the doctoral and master's students for their hard work and dedication!

Thanks to Southeast University Press for the support of the publication of this book, especially the editor Xia Lili. She is serious and responsible, and helped to improve the quality of the manuscripts. I would like to express my heartfelt thanks to her hard work!

The research work in this book is mostly supported by the National Natural Science Foundation projects "Modeling and Precise Motion Control of Tendon-Sheath-Driven Soft and Slender Robot" (50875044), "Soft Operation Technology of Tendon-Sheath-Driven Soft and Slender Robot" (51175078), "Realization of artificial muscle based on Tendon-Sheath and its application in wearable rehabilitation devices" (51575100), and "Slender robots based on multi-tendon-sheath transmission and variable stiffness technology" (52175005), and also partly supported by of Jiangsu Provincial Natural Science Foundation project "Modeling and Control of Tendon-Sheath-Driven Soft Slender Detection and Rescue Robot" (BK2008310). And presently, we are supported by Jiangsu Province key research and development project: "Development of Key Technologies of slender robot for detection of narrow space inside Special equipment" (BE2022153).

Nanjing, China
December 2024

Xingsong Wang

Competing Interests The authors have no competing interests to declare that are relevant to the content of this manuscript.

Contents

Chapter 1
Introduction

Abstract The tendon-sheath mechanism is an important method of mechanical transmission. Different from gears, worm gears and other transmission methods, tendon-sheath transmission has outstanding advantages like lightness and flexible transmission path. Due to these benefits, it has been widely used in various robots, including slender robots, medical robots, exoskeleton robots, etc. In this chapter, we discuss slender robots and delve into how tendon-sheath transmission is applied in these systems. Additionally, the current research of tendon-sheath transmission theory will also be discussed. Moreover, the bionic prospect of tendon-sheath design is also explored with the inspiration of skeletal muscles. This chapter will help readers to understand the tendon-sheath transmission and its main robot application status.

1.1 Introduction

Elongated robots, are specialized bionic robots designed with a snake-like function. They are essential for tasks like search and rescue during natural disasters or accidents, especially in complex and tight spaces. These robots are useful for inspecting buildings, electrical systems, heating pipelines, and even performing surgeries in human cavities. Over the years, research has advanced in their design and control, leading to significant progress.

In terms of search and rescue robots, current robots for search and rescue inside disaster ruins generally have the disadvantages of large size, inflexible operation and poor adaptability to the environment. This poses a significant problem, especially when considering the critical "golden 72 h" window where finding and rescuing trapped individuals becomes increasingly difficult. Therefore, it is very important to develop robots that can navigate deep inside the rubble to improve the rescue rate of people trapped inside the rubble. To address this, slender robots emerge as a promising solution, it can operate effectively in curved and narrow spaces, as Fig. 1.1 shows. With the slender robot technology, we can substantially elevate the capabilities of search and rescue missions within ruins.

On the other hand, the large equipment maintenance(such as large transformers, aircraft wings, aircraft engines, large turbines, etc.) mostly depends on traditional

X. Wang and Q. Zhang, *Tendon-Sheath Transmitted Light and Slender Robots*,
https://doi.org/10.1007/978-981-96-5581-6_1

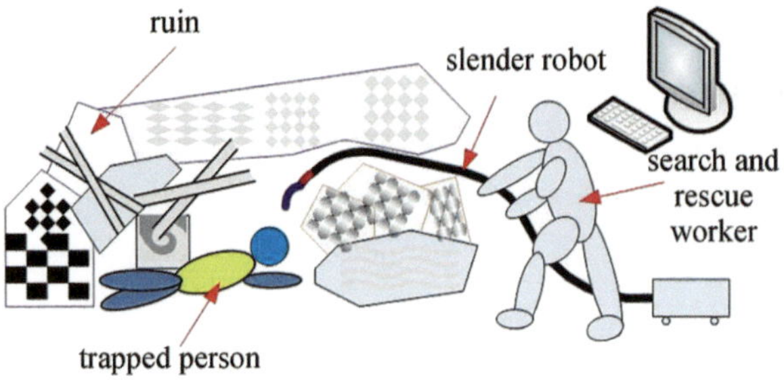

Fig. 1.1 Search and rescue inside the ruins of the disaster

methods like preventive experiments PT (Prevention Test) and periodic maintenance TBM (Time Based Maintenance). As the demand for advanced equipment operation and maintenance is growing rapidly, present limitations associates with these methods are high costs, operational blind spots, and inconvenient disassembly. To this end, elongated snake robots prove invaluable for inspecting and maintaining high-end equipment, given their ability to navigate tight spaces and access critical components, as shown in Fig. 1.2. Thus, using slender robots for non-disruptive internal equipment maintenance has gained traction. This approach promises enhanced maintenance efficiency and post-maintenance equipment reliability.

Additionally, minimally invasive surgical techniques, emerging in response to the limitations of traditional surgery, have significantly revolutionized the medical field in recent decades. This type of surgery makes one or more small incisions or uses natural body openings for operations, and guided by visuals. These less invasive techniques are gaining popularity due to reduced patient discomforts, faster postoperative recovery and cost-effectiveness. Currently, robots designed for minimally invasive procedures are garnering significant global attention. Several research findings have successfully transitioned from the lab to clinical applications, driving commercialization efforts.

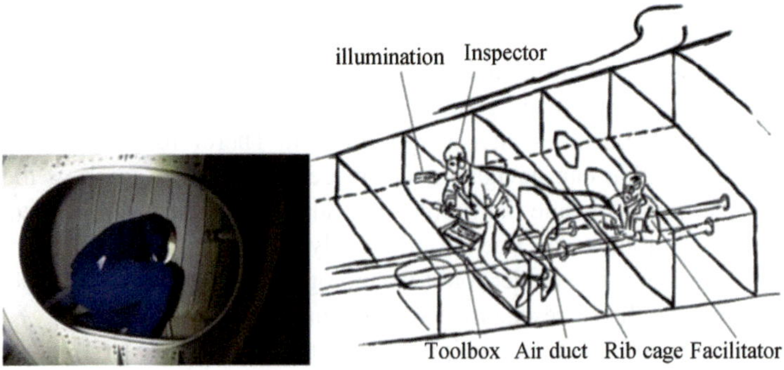

Fig. 1.2 Inspection crews enter the wing tanks to carry out inspection operations

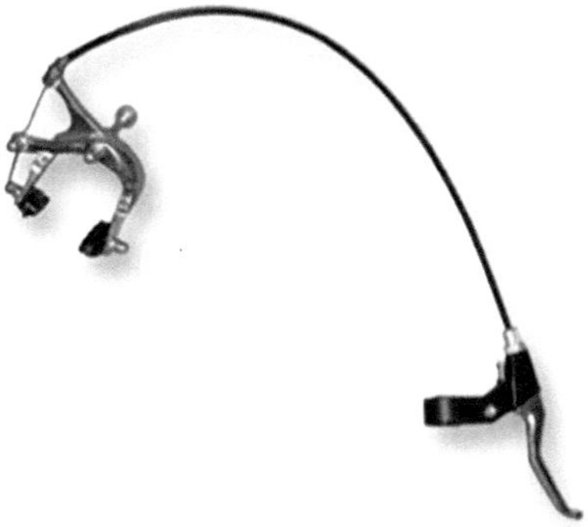

Fig. 1.3 Tendon-sheath brake cable

The main component of minimally invasive surgical robots is a slender manipulator that enters the body through small incisions or natural openings. Its size, flexibility, and precision directly influence the quality and recovery time of the surgery. Therefore, the advancement of slender robot technology directly affects the effectiveness of minimally invasive surgical procedures.

Tendon-sheath transmission is widely used in brake and control systems for bicycles, motorbikes due to its lightweight, compact size, and flexibility. This system comprises a slender "rope" encased within a "sleeve," serving as a primary method for flexible, long-distance remote transmission. A notable application is in bicycle brake cables (Fig. 1.3). Given its soft structure, it can navigate narrow spaces and cover extended distances, such that offering simplicity in mechanical design. Owing to these benefits, tendon-sheath transmissions have gained prominence in designing robotic hands, minimally invasive surgical robots, and continuum robots in recent years.

1.2 Light and Slender Robots

1.2.1 Status of Research on Slender Search and Rescue Robots

The slender robot is a new type of special robot, mainly used in non-structural environments. According to the deformed method of the slender body, it can be divided into full-length deformation controllable and partially controllable and partially passive.

(1) Full-length deformation controllable slender robots: This type of robot is mainly a class of bionic continuum robots, which imitate the movement mechanism of animal organs such as elephant trunk and octopus in nature, and do not have

movement joints themselves, but can rely on continuous flexible deformation to achieve movement and grasping operations.

(2) Partly controllable, partly passive slender robots: This type of robot is mainly a class of endoscopic surgical robots that penetrate deep inside the human body (stomach, colon, etc.) through the innate tubular cavities (nose, mouth, etc.) to perform diagnostic, therapeutic, testing and surgical operations. They generally have a passively curved body that can pass through the narrow curved intestines of the body and a head with multiple controlled degrees of freedom to perform various operations.

These slender robots utilize a variety of locomotion methods to adapt to different environments and challenges [1]. The primary forms of movement include wheeled, tracked, pneumatic, deformable material based mechanisms. Each type of locomotion is designed to address specific challenges in search and rescue operations. By utilizing the appropriate movement form, these robots can efficiently navigate diverse environments, providing critical support in disaster response and industrial maintenance.

A snake-like robot, Souryu, for post-disaster debris search and rescue, was designed by K. Osuka et al. at the International Rescue Systems Research Institute in Japan [2]. The robot is actually a tracked cart driven by an electric motor, it uses 12 cameras and a panoramic camera mounted on the body to acquire real-time images. To determine the robot's posture in the rubble, a Flexible Sensor Tube [3] consisting of several hollow joints in series was designed. Each joint is equipped with a potentiometer to measure the relative angle between the joints, and the sensor data can be used to determine the shape of the flexible tube and the position of the vehicle.

What's more, a slender mobile robot for search and rescue within debris was developed at Shanghai Jiao Tong University, it has four main components: a front-end multi-sensor unit, a motor unit, a propulsion unit, and a passive unit. The front-end multi-sensor unit is capable of data and network transmission of video, audio, temperature and CO_2 concentration. In order to reduce the diameter of the robot, the propulsion unit and the motor unit are separated from each other and are connected to each other by a universal drive shaft. The propulsion unit has 10 reciprocating propellers evenly distributed around the circumference, with the adjacent propellers moving alternately to achieve continuous propulsion of the robot. There are two attitude drive shafts between the motor unit and the propulsion unit to adjust the attitude of the propulsion unit with respect to the motor unit. The passive unit is a slender multi-degree-of-freedom multi-sensor hose mechanism with angular encoders, which enables three-dimensional positioning and shape reconstruction of the robot through the positive kinematics of the passive segment. The robot was tested under laboratory conditions with a maximum crawling gradient of approximately 23.42, a maximum turning angle range of –40 to +40 at the front end of the robot and a maximum crawling speed of 5 cm/s.

Tohoku University developed an Air-floating Snake-like Climbing Robot (ASC) [4], its body is covered with inclined cilia, and driven by a vibration motor. The tip is controlled by an air nozzle, allowing for height and directional movement. It is 7 m long, 50 mm in diameter, weighs 3 kg, has a camera at the tip, and has nylon cilia that are 8 mm long and 0.2 mm thick. Its tip is pneumatically driven, allowing the floating height to be adjusted via controlled air pressure, while the nozzle can rotate within a 50 range to change tip orientation. The ASC robot has good mobility both on flat ground and over obstacles.

In order to enhance the mobility of the snake robot, and overcome the load-bearing obstacles caused by the slender body, a driving assistance mechanism (DAM) is proposed to assist a snake-like robot in moving on steep slopes [5]. The DAM is a passive assisting mobility structure that bears the weight of the snake robot. Therefore, when the snake robot moves sideways on a steep slope, the DAM can help prevent it from rolling over. Such that the driving performance of the snake robot on steep slopes is significantly improved in this design.

In conclusion, the current search and rescue robots are generally oversized due to the use of pneumatic or motor direct drives, with most robots integrating drive elements, various sensors and therefore existing slender search and rescue robots generally have the disadvantage of being oversized. The large size of the robot body causes some difficulty in getting the robot into small spaces. The robots generally have fewer degrees of freedom to control, which makes them less flexible and their end posture less easily adjustable. Although some smart materials such as SMA can reduce the weight of the driving mechanism and obtain better flexibility, their slow response speed and insufficient driving capability limit their applications [6]. In view of the complexity of the working environment, the majority of robots are controlled by the head only and the body is flexible and passive in order to reduce the complexity of the robot structure and control.

1.2.2 State of the Art on Continuum Robotics

Continuum robots do not contain any rotating joints or rigid links and are similar to snake bodies, elephant trunks and octopus tentacles in that they can bend continuously along their length through elastic deformation [7].

Walker and Jones at Clemson University have investigated the structure, kinematics and dynamics of continuum robots in a variety of ways [8]. They designed various robots with different continuum skeletons: a stainless steel bar skeleton, a two-degree-of-freedom hinge skeleton, a pneumatic muscle skeleton and an inflatable flexible pipe skeleton. The robot is divided into segments along its length, and the deformation of each length except the pneumatic muscles is controlled by cords arranged on the outside of the robot, with three or four cords arranged in the circumferential direction of the robot. The 'elephant trunk' robot, consisting of three pneumatic muscles side by side, relies on the length of the pneumatic muscles to bend and deform. They used differential geometry to derive and simulate the kinetic equations of a continuum robot and proposed a method to obtain the shape of the robot in real time using machine vision. For a class of multi-segment tandem continuum robots that rely on rope length variation for body deformation, they derived their general kinematic equations to obtain the correspondence between the length, curvature and bending angle of robot deformation and rope length variation [9] .

The most successfully commercialised continuous robot is the snake-arm "Snake-Arm" developed by OC Robotics. Similar to the spring-frame robot from Hanyang University in Korea, it has a hollow flexible body as its backbone and three ropes arranged evenly around its circumference to bend the robot. The individual bending joints are of modular design and can be increased in length as required. The entire snake robot can be mounted at the end of a Kuka industrial robot as an end operator, thus enabling the snake robot to enter tight spaces inside the wing that are inaccessible to workers, and to perform tasks such as internal monitoring/standard fastening and gluing, as well as walking/threading, chip and liquid suction, leak monitoring, spray weld tracking and inspection, deburring of aircraft components etc. [12]. As a non-loop-breaking inspection and installation technique, the use of a serpentine robot eliminates the need to disassemble and reassemble aircraft parts. With devices such as cameras and grippers mounted on the end of the snake robot, it can be used for nuclear reactor pipeline access.

Today, continuum robots have been developed in a variety of configurations, in addition to the elastic skeleton described above, the pneumatic or hydraulic actuator is also an effective way to construct the continuum robots. Tianjin University has developed a variable stiffness continuum robot for flexible gastrointestinal endoscopy applications, it is driven by a hybrid of pneumatic and cable systems. Based on this hybrid driving method, the continuum robot has obtained flexible bending path adaptability, while having sufficient stiffness and load force to cope with surgical operations [10].

In addition, the concentric tube continuum robot has also shown good performance in applications with finer dimensions, such as the vascular intervention [11]. However, the unique elastic deformation structure of concentric tubes and the high-precision operation requirements in complex human environments have made their motion control and planning a critical issue. Girerd etc. studied the automatic navigation and obstacle avoidance issues of concentric tube robots in positional environments and proposed an internal automatic navigation method [12]. what's more, the deformation structure based on smart materials makes the driving and sensing structures of the continuum more diverse, such as shape memory alloy (SMA).

From the current state of research on continuum robots it can be seen that the controllability of the whole body of the continuum robot gives it good bending performance and shows good potential for applications in industry and minimally invasive surgery. The well-defined kinematic relationships between the bending and deforming joints of the continuum robot ensure its path planning within a known environment, but the presence of some coupling between the movements of the joints makes it less adaptable to unknown unstructured environments.

1.3 Tendon-Sheath Transmission

1.3.1 State of the Art on Modelling and Control of Tendon-Sheath Drive Systems

The friction between the cable and the sheath, combined with the variation in rope length, results in significant nonlinearity in displacement and tension/torque transmission, which affects the precision application of the tendon-sheath system. Therefore, it is of great significance to model the tendon-sheath drive system, study its transmission characteristics and obtain a method for its compensation control.

The first theoretical analysis of the transmission characteristics of the tendon-sheath was carried out by the Japanese scholar Okada when he designed the Okada Hand. However, they only noticed the effect of friction between the rope and the sleeve on the gripping force of the manipulator and neglected the error caused by the deformation of the rope on the position control of the manipulator. Subsequently, another Japanese scholar, M. Kaneko, considered the transfer characteristics of the tendon-sheath force and position together. They developed a static model for the transfer of force and position of a single tendon-sheath under constant preload based on the Coulomb friction model and demonstrated the correctness of the model through computer simulations and experimental results. In the Darmstadt Hand torque control experiments they found that the force and position transfer in a double-cable coupled transmission exhibited properties similar to both mechanical gap and hysteresis.

Based on the Coulomb model, Professor B. Yao at Purdue University first described a fully continuous transmission model in the time domain for any preload force and sheath curvature using a set of differential equations [13], which described both force and position transfer characteristics. They then extended the model to the case of a double-cable coupled drive. Simulation and experimental results showed that there is interaction between the two ropes in a double-cable coupled drive resulting in a 'spike' in force transfer from a single cable that is not present in a purely single-cable drive. They proposed an inverse mode compensation method with online estimation of parameters for the gap-like phenomenon that occurs in dual-cable drives. Taking into account the performance limitations of the drive, they designed a smooth inverse model of the gap using a second-order filter to avoid jitter in the drive output. Experimental results of this model on a two-cable test stand and on the Laprotek surgical robot show that the method is effective in reducing the transmission error in position.

However, the Coulomb friction model cannot describe the dynamic transmission characteristics under high-speed motion. Therefore, more dynamic models based on the Bouc-Wen model, the Lugre friction model, the Preisach model, and the Dahl model have been proposed to describe the hysteresis characteristics of the tendon-sheath mechanism. Do et al. have carried out research on the tendon-sheath transmission model based on the Bouc-Wen model and Dahl model. They designed a feedforward controller based on the transmission model to compensate the hysteresis loss, and used genetic algorithm to identify the model parameters [14]. In this paper, a simplified piecewise linear model is proposed to describe the gap hysteresis and establish the dead-zone compensator. In addition, the motor currents from the TSM robot controller are used to identify the model parameters. Lee D H et al. proposes a simplified piecewise linear model to describe the hysteresis of tendon-sheath transmission, and establish a dead-zone compensation controller. In addition, they use the motor currents from the driving end to identify the model parameters, this method simplifies the calculation process and is suitable for practical applications [15].

With the development of artificial intelligence algorithms, new methods with machine learning have emerged for the establishment and output prediction of nonlinear systems. Kim D. proposed a hysteresis model based on the Preisach hysteresis model and a recurrent neural network (RNN), it can predict the transmission output with high accuracy as the deformation property of the sheath was considered [16]. In addition, many controllers and prediction of nonlinear transmission characteristics also incorporate deep learning-based approaches. However, their parameter identification requires a large amount of training data and computation, and the complex structure of these models will make real-time control more difficult.

1.3.2 Status of Research on Tendon-Sheath Drive Mechanisms

Tendon-sheath mechanisms are widely used in the design of various detection robots, minimally invasive surgical robots, robotic dexterous hands and bionic exoskeletons because of their soft structure, long transmission distances, small size, light weight and lack of need for transition pulleys.

The da Vinci surgical robot is a typical example of using tendon-sheath mechanism to perform delicate surgical operations. Unlike the Da Vinci and Laprotek surgical robots, which require access to the body through small holes cut in the human body, Phee et al. of Nanyang Polytechnic, Singapore, designed a 22 mm diameter master-slave endoscopic robot [17] that works through a surgical jaw channel on an Olympus 2T610 endoscope. This robot is also known as a NOTES (Natural orifice transluminal endoscopic surgery) surgical robot because it does not require a cut in the patient. The robot has nine degrees of freedom in both arms, each driven independently by a pair of lassoes. The movements of the operator's hand are mapped directly to the end hand via the tendon-sheath drive. Imperial College London proposes the LaryngoTORS robotic system for the Transoral Laser Phonosurgery. It controls the laser fiber tip through a cable-driven parallel mechanism (CDPM), the utilize of the parallel mechanism have achieved high motion accuracy and good repeatability [18]. What's more, G. Palli et al. from the University of Bologna, Italy, using the tendon-sheath mechanism to develop the UB Hand IV and achieve flexible human-like grasping and gesture movement [19]. Due to the large number of hand joints and the high flexibility and driving force requirements, the tendon-sheath system has become the main transmission method for current humanoid hand designs.

The tendon-sheath mechanism is also widely used in exoskeletons because of its softness and flexibility. It can easily adapt to the surface of human body and separate the drive device and the execution unit. With the use of tendon-sheath mechanism, the entire exoskeleton structure is lightweight and has good human compatibility [20]. Collins's group in Stanford University has designed an untethered ankle exoskeleton that includes an ankle assist device and a waist battery pack [21]. The whole machine has low mass and can provide a peak torque of 54 Nm, which is about 50–75% of the human ankle torque. To achieve outdoor adaptive walking, the exoskeleton-assisted motion parameters were optimized based on a data-driven approach, and a speed adaptive controller was developed that can adjust the exoskeleton motion assistance according to the walking speed. In addition, the tendon-sheath mechanisms have also been widely used in wearable gloves, soft robots, flexible manipulation grippers, etc. due to its arbitrary bending path and good flexibility [22, 23].

1.4 Bio-aspects of Tendon-Sheath Transmission

The movements of biological joints are generated in a quite compliant, energy-efficient and adaptive way owing to the special characteristics of muscles, e.g., high energy density, low specific gravity, and compliant structure. However, it is another case in the mechanical joints of robots which are driven by traditional actuators with hard, heavy, and power-consuming structures. Although traditional actuators can accurately track the torque or displacement trajectory, their hard shape and invariant stiffness will give rise to unnecessary impulse or energy consumption, even safety problems especially in the man-machine interaction designs. With increasing demands for human-robot cooperation, good performance and insured safety have become important considerations for actuator design.

Over the past few decades, a lot of efforts have been devoted to develop compliant actuators. The biologically inspired elastic actuators, such as series elastic actuator, parallel elastic actuator, series-parallel elastic actuator, and variable stiffness actuator, demonstrate improvement in safe and comfortable actuation. These elastic actuators are compliant, energy-efficient, and shock absorbing, nevertheless the complex mechanism designs usually come with large size, heavy weight, or hard structure, therefore, they are different from human muscle. In order to improve the biological similarity, a large number of actuators as known as artificial muscles are developed. These artificial muscles that are made of shape memory alloys, dielectric elastomer, or carbon nanotubes have controllable contractions, high energy density ratios, and soft structures. They have shown the great application potential in the field of bionic robots. However, due to the deformation properties and the driving limitations, these artificial muscles are currently not suitable for the wearable exoskeleton which requires good portability, large output force, and precise movement control. In addition to these mentioned shortages, current artificial muscles also lack muscle contraction property. The unique active contraction property is essential for the biological muscle to be the best natural actuator to adapt and balance the external loads.

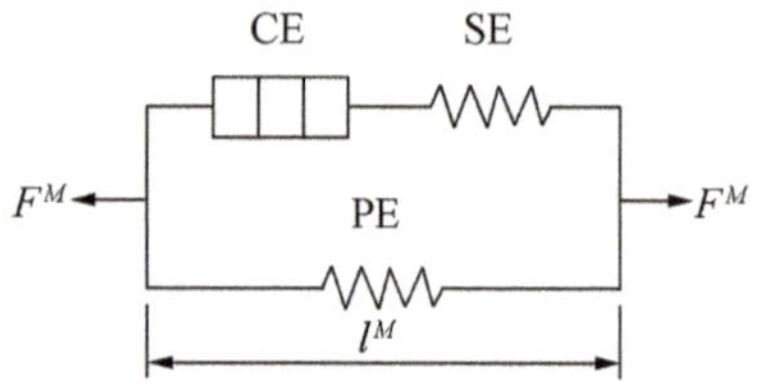

Fig. 1.4 Hill three-element muscle model. Reprinted from [24] with permission from IEEE

Muscles are composed of fiber bundles that consist of numbers of muscle fibers. These fibers contain many myofibrils which are correspondingly composed of many sarcomeres. Muscle contraction occurs when sarcomeres are stimulated by nerve impulse. The complex structure and contraction mechanism make muscle elasticity and force adjusted in a natural manner. Muscles cannot contract actively in the absence of stimulation, nevertheless they can still withstand passive stretching to provide deceleration and protection for joint motions, owing to the elasticity of the tendon and myofilaments. The mechanical properties of muscles can be expressed by muscle models, in which Hill muscle model are commonly used. It is a phenomenological model, it expresses the mechanical properties of muscle based on experimental data. Specifically, the Hill muscle model includes three basic elements, as shown in Fig. 1.4: CE represents the active fiber bundles which contribute to the active contractile force of muscle, it is activated by stimulation, and the contractile force is nonlinearly related to the length and velocity of muscle; SE represents the elasticity of tendon and myofilaments, the force-extension relation of SE is commonly modeled as quadratic; PE represents the muscle elasticity without activation, it is generally believed that the relationship between force and extension is similar to that of SE.

Inspired by Hill muscle model, an artificial muscle with the three elements is developed [24], as Fig. 1.5 shows. First of all, in order to provide active contraction, and ensure that the artificial muscle has small size and light weight, a tendon-sheath actuation system is utilized as CE. SE and PE are modeled as nonlinear elasticities in Hill muscle model, but they are all replaced by linear springs for simplicity. The simplification makes sense on account of the fact that the sharp increased stiffness of the two elastic elements occur in the case of ultimate tension, whereas the artificial

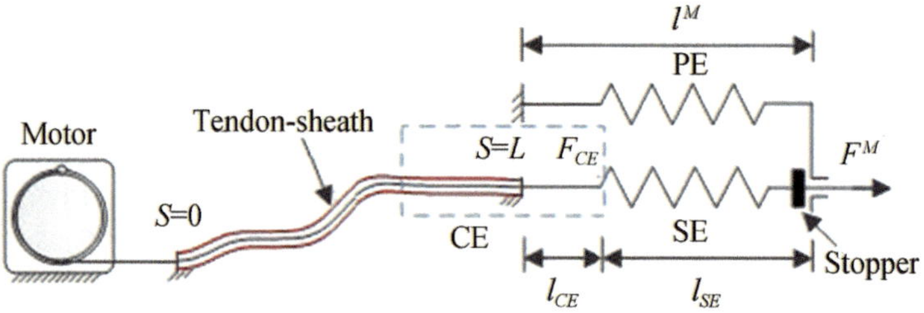

Fig. 1.5 The schematic diagram of tendon-sheath artificial muscle. Reprinted from [24] with permission from IEEE

muscle mainly works under normal contraction length. What's more, it will simplify the modeling as well as control of the artificial muscle.

The length of artificial muscle l^M can be taken as $l^M = l_{CE} + l_{SE}$. To achieve the passive tensile properties of muscle, a stopper is provided between SE and PE. When $l^M > l_0{}^M$ ($l_0{}^M$ is the length of the artificial muscle in original state), the muscle is stretching and the parallel spring is stretched by the stopper, otherwise the parallel spring will retain the original length. As a result, the tendon-sheath artificial muscle is similar to Hill muscle model. The elastic behavior of the artificial muscle can be described by the equivalent stiffness which is defined as Eq. (1.4.1).

$$K^M = \frac{K_{SI} K_{CE}}{K_{SE} + K_{CE}} + K_{PE} \tag{1.4.1}$$

As Fig. 1.6 shows, K^M has a nonlinear relationship with abs(K_{CE}). When abs(K_{CE}) is relatively low, K^M is mainly determined by CE. However, as the abs(K_{CE}) increases, K^M is gradually approaching to $K_{CE} + K_{SE}$ instead of infinitely increasing, so a more compliant artificial muscle is brought out. In the case of zero active contraction, $K^M = K_{PE}$, which represents the elasticity of relaxed muscle. The stiffness of the tendon-sheath artificial muscle is variable according to CE and limited by the passive elasticities, which is similar to muscle elasticity. In general, the tendon-sheath artificial muscle has compliant structure, high power transmission capacity, variable elasticity, contraction modes similar to biological muscle. It also proves the great potential of tendon-sheath transmission in bionic drive system.

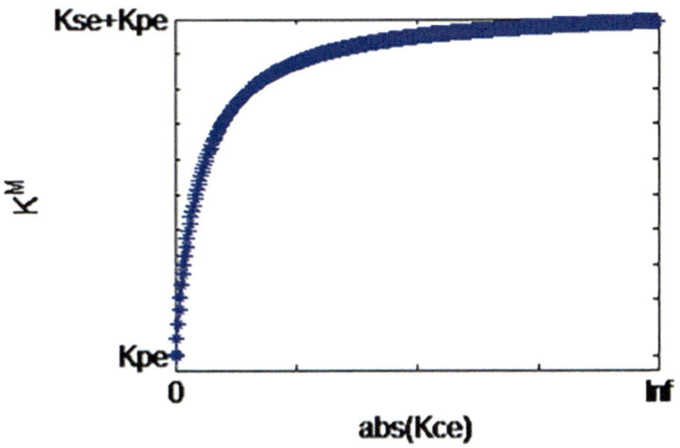

Fig. 1.6 The relationship between the equivalent stiffness (K^M) and the absolute active equivalent stiffness (abs(K_{CE})) of the tendon-sheath artificial muscle. Reprinted from [24] with permission from IEEE

References

1. Liu J, Tong Y, Liu J (2021) Review of snake robots in constrained environments. Robot Auton Syst 141:103785
2. Arai M, Tanaka Y, Hirose S et al (2008) Development of "Souryu-IV" and "Souryu-V": serially connected crawler vehicles for in-rubble searching operations. J Field Robot 25(1):31–65
3. Osuka K, Doi T, Tadokoro S et al (2006) In-rubble robot system for USAR under debris. SICE-ICASE international joint conference, pp 3439–3442
4. Yamauchi Y, Fujimoto T, Ishii A et al (2018) A robotic thruster that can handle hairy flexible cable of serpentine robots for disaster inspection. In: 2018 IEEE/ASME international conference on advanced intelligent mechatronics (AIM). IEEE, pp 107–113
5. Bae J, Kim M, Song B et al (2020) Snake robot with driving assistant mechanism. Appl Sci 10(21):7478
6. Cortez R, Sandoval-Chileo M, Lozada-Castillo N et al (2024) Snake robot with motion based on shape memory alloy spring-shaped actuators. Biomimetics 9(3):180
7. Russo M, Sadati S, Dong X et al (2023) Continuum robots: an overview. Adv Intell Syst 5(5):2200367
8. Gravagne I, Rahn C, Walker I (2003) Large deflection dynamics and control for planar continuum robots. IEEE/ASME Trans Mechatron 8(2):299–307
9. Jones B, Walker I (2006) Kinematics for multisection continuum robots. IEEE Trans Robot 22(1):43–55
10. Luo X, Song D, Zhang Z et al (2023) A novel distal hybrid pneumatic/cable-driven continuum joint with variable stiffness capacity for flexible gastrointestinal endoscopy. Adv Intell Syst 5(6):2200403
11. Alfalahi H, Renda F, Stefanini C (2020) Concentric tube robots for minimally invasive surgery: current applications and future opportunities. IEEE Trans Med Robot Bionics 2(3):410–424
12. Girerd C, Kudryavtsev A, Rougeot P et al (2020) SLAM-based follow-the-leader deployment of concentric tube robots. IEEE Robot Autom Lett 5(2):548–555
13. Agrawal V, Peine W, Yao B (2010) Modeling of transmission characteristics across a cable-conduit system. IEEE Trans Robot 26(5):914–924
14. Do T, Tjahjowidodo T, Lau M et al (2016) Real-time enhancement of tracking performances for cable-conduit mechanisms-driven flexible robots. Robot Comput-Integr Manuf 37:197–207
15. Lee D, Kim Y, Collins J et al (2021) Non-linear hysteresis compensation of a tendon-sheath-driven robotic manipulator using motor current. IEEE Robot Autom Lett 6(2):1224–1231
16. Kim D, Kim H, Jin S (2022) Recurrent neural network with preisach model for configuration-specific hysteresis modeling of tendon-sheath mechanism. IEEE Robot Autom Lett 7(2):2763–2770
17. Phee L, Low S, Thant Z et al (2008) Robotic system for no-scar gastrointestinal surgery. Int J Med Robot Comput Assist Surg 4:15–22
18. Zhao M, Vrielink T, Kogkas A et al (2020) LaryngoTORS: a novel cable-driven parallel robotic system for transoral laser phonosurgery. IEEE Robot Autom Lett 5(2):1516–1523
19. Melchiorri C, Palli G, Berselli G et al (2013) Development of the UB hand IV: overview of design solutions and enabling technologies. IEEE Robot Autom Mag 20(3):72–81
20. Deman M, Asfour T, Ude A et al (2022) Mechanical design and friction modelling of a cable-driven upper-limb exoskeleton. Mech Mach Theory 171:104746
21. Slade P, Kochenderfer M, Delp S et al (2022) Personalizing exoskeleton assistance while walking in the real world. Nature 610(7931):277–282
22. Xiong H, Diao X (2020) A review of cable-driven rehabilitation devices. Disabil Rehabil: Assist Technol 15(8):885–897
23. Zaidi S, Maselli M, Laschi C et al (2021) Actuation technologies for soft robot grippers and manipulators: a review. Curr Robot Rep 2(3):355–369
24. Zhang Q, Wang X, Tian M et al (2017) Modeling of novel compound tendon-sheath artificial muscle inspired by hill muscle model. IEEE Trans Ind Electron 65(8):6372–6381

Chapter 2
Tendon-Sheath Transmission System

Abstract Brake lines are a classic application of the tendon-sheath actuation system. This lightweight, cost-effective, and simple actuation method, also known as the tendon-sheath or Bowden cable system, is widely utilized in various applications such as dexterous robotic hands, Minimally Invasive Surgery (MIS) systems, and exoskeleton robots. In practical applications, there are two primary configurations of the tendon-sheath actuation mechanism: single tendon-sheath transmission and double tendon-sheath transmission. Furthermore, for more flexible, lightweight, small-sized, and cost-effective robotic applications, we propose two types of multi-tendon-sheath transmission mechanisms: one utilizing a multi-channel sheath and another driven by a single motor. This chapter will introduce these tendon-sheath transmission forms and discuss their characteristics, providing guidance for selecting the appropriate tendon-sheath transmission forms in different robotic applications.

2.1 Single Tendon-Sheath Transmission System

The tendon-sheath transmission, illustrated in Fig. 2.1, is a cable transmission mechanism comprising an inner tendon and a flexible outer sheath. In practical applications, the two ends of the outer sheath are securely fixed, allowing the sheath to bend as needed. The inner tendon moves within the sheath to convey force and displacement from one end to the other. The tendon-sheath system is typically powered by a motor, the inner tendon is activated when pulled by the motor. As the tensioned tendon slides within the sheath, it transmits both movement and force to the opposite end. Arrows on the tendon indicate the direction of movement.

The outer sheath features a slender body with a smooth inner wall, exhibiting flexibility and axial incompressibility. This allows it to bend freely, adapting to the desired transmission path. The inner tendon typically consists of a stainless steel wire rope, slightly smaller in diameter than the inner diameter of the sheath. Both the tendon and outer sheath are readily accessible from commercial producers, making them affordable and widely available. As Fig. 2.2 shows, the brake line that consists of stainless steel wire rope and brake line pipe is a classical tendon-sheath mechanism, the tendon usually with a diameter of 1.6–3 mm, can transfer force up to thousands

X. Wang and Q. Zhang, *Tendon-Sheath Transmitted Light and Slender Robots*,
https://doi.org/10.1007/978-981-96-5581-6_2

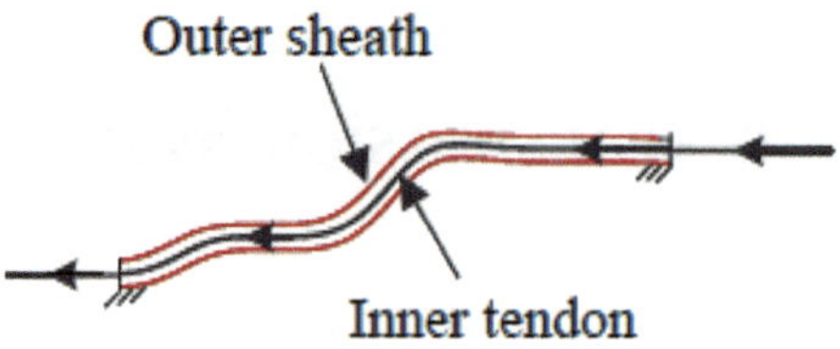

Fig. 2.1 The schematic diagram of tendon-sheath transmission mechanism. Reprinted from [24] in Chap. 1 with permission from IEEE

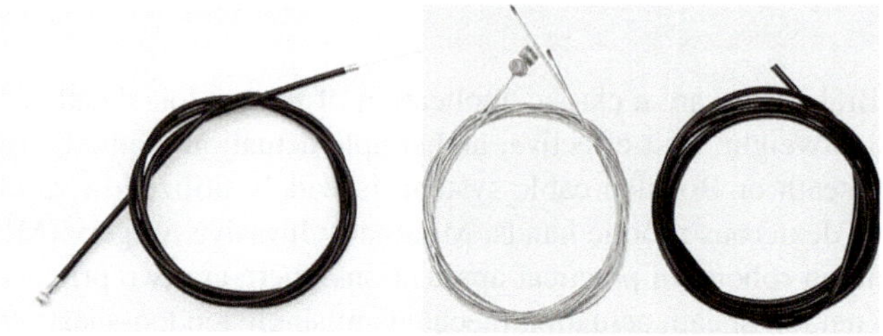

(a) The brake line consists of steel wire rope and line tube

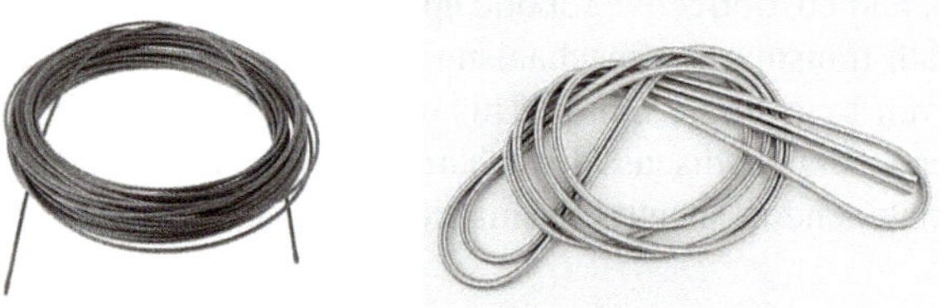

(b) Thin wire rope (c) Spring sheath

Fig. 2.2 The commonly used tendon and sheath

of Newton. This mechanism can be used in robotics systems requiring high loads, such as exoskeletons. The wire rope with smaller diameter (usually 0.5–1.2 mm) in Fig. 2.2b can be used on small robotic systems with low loads. Accompanied by the thin wire rope, we usually take tight spiral springs as the outer sheath, shown in Fig. 2.2c, it is smaller and lighter. Such combination of tendon-sheath can be used in smaller or slender robots like endoscopy robots.

The tendon-sheath transmission mechanism is simple, compact, cost-effective, and easily bendable. It operates with flexibility, allowing motion transmission through narrow or curved paths. Consequently, this mechanism has been widely adopted in robotic applications [1, 2]. There are two primary forms of tendon-sheath transmission: the single tendon-sheath transmission and double tendon-sheath transmission.

Figure 2.3 shows the configuration of a single tendon-sheath transmission mechanism for linear motion joint. In this setup, motion from the driving end is conveyed to the linear joint through the tendon-sheath mechanism, propelling the driven joint forward. The number of tendons (stainless steel cables) may vary based on specific

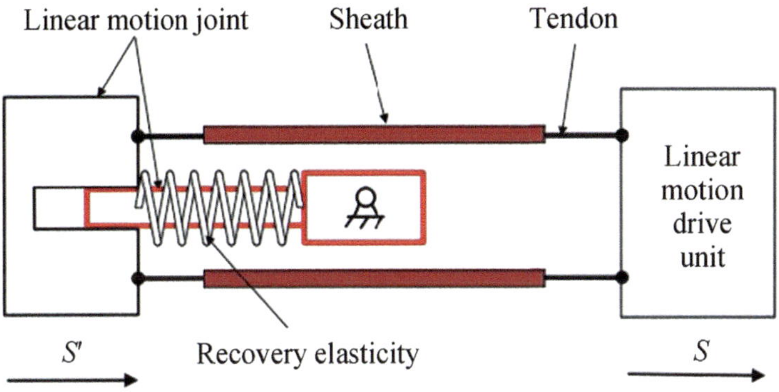

Fig. 2.3 Single tendon-sheath transmission mechanism for linear joint

circumstances, with each cable operating independently during transmission. However, as cables can only stretch and not compress, it is crucial to include corresponding elastic recovery elements when utilizing cables for mobile joint actuation. This incorporation ensures that when the force acting on the cables gradually decreases, the driven component of the mobile joint can return to its original position with the assistance of the reactive force from the elastic element.

As a result, in the cable transmission process, the forward movement of the joint's driving component is controlled, allowing the cables to convey the input to the driven component and propel it forward. Conversely, when the force on the driving component decreases, the driven component can use the reactive force of the elastic element to return to its original position.

For rotary joints, the single tendon-sheath transmission mechanism can be configured as Fig. 2.4 shows. One pulley is active while the other is passive. In this driving mode, the joint is propelled in one direction by a single active pulley, which can be powered directly by a motor or operated manually. Since the tendon only sustains

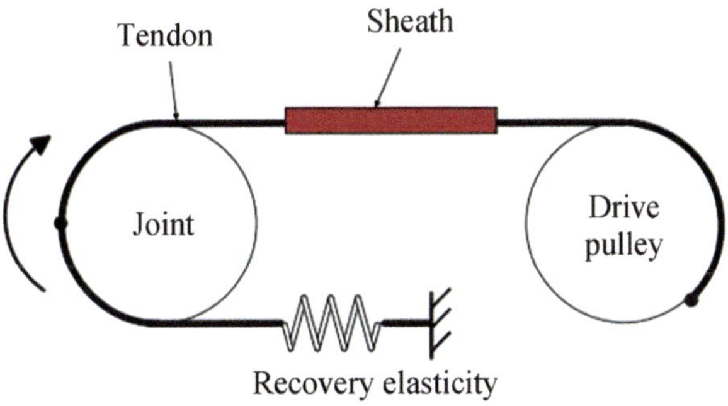

Fig. 2.4 Single tendon-sheath transmission mechanism for rotary joint

tension, the joint's motion in the opposite direction is under-actuated, and an elastic component, such as a spring, is used to store energy and pull the joint back.

The single tendon-sheath transmission is very suitable for lightweight and highly integrated design of multi-degree-of-freedom mechanisms [3, 4]. A dexterous bionic hand is developed by multiple sets of tendon-sheath transmission mechanisms, with single tendon-sheath mechanisms and small springs in the tiny joints of fingers, the robot hand owns 19 degrees of freedom [5]. The Exo-Glove Poly (EGP) II is a wearable rehabilitation glove, it is made of polymer material and driven by the single tendon-sheath mechanism, the overall structure is soft and lightweight [6]. In addition, some wearable exoskeletons are also driven by a single tendon-sheath transmission mechanism [7, 8]. Harvard University designed a new generation of lightweight and flexible wearable power-assisted exoskeleton using tendon-sheath transmission mechanisms [9], the Soft Exosuits. Different from traditional rigid exoskeletons, this exoskeleton is as soft and light as clothes.

2.2 Double Tendon-Sheath Transmission System

Figure 2.5 illustrates the setup of a double tendon-sheath transmission mechanism, featuring two tendons placed antagonistically. This configuration enables the joint to move in both directions, as each tendon independently actuates the joint [10]. To prevent interference between the tendons, one must be loosened when the other is in action. Consequently, only one cable drives the end joint at any given moment, leveraging the transmission characteristics of a single cable for effective joint control. This method is commonly applied in scenarios demanding precision in position and force control. However, employing this transmission mode requires two drive units for a joint to achieve independent control in two directions, introducing a level of operational complexity.

Figure 2.6 displays the configuration of a double tendon-sheath transmission mechanism with a coupled antagonistic setting. The end joint in this system is actuated by two coupled tendons, requiring only one motor for operation. To prevent tendon slipping, the ends of each tendon are fixed to the pulley, as depicted in Fig. 2.6. This setup allows a 180° rotary motion of the end joint. Introducing a multi-turn spiral guide groove on the cylindrical surface of the rotary pulleys can expand the rotation range and enable multi-turn movement. However, as the tendons are fixed, the tendon-sheath mechanism cannot execute continuous movement like a belt transmission.

The double tendon-sheath transmission facilitates two-way rotary motion. Particularly useful in slender and intricate operating spaces, it permits the separation of the drive end and the end effector. This separation allows for the remote drive of the end operating mechanism, leading to a reduction in the size of the operating mechanism. Consequently, this enhances flexibility and enables the mechanism to function effectively within limited spaces. Due to these advantages, it finds wide application

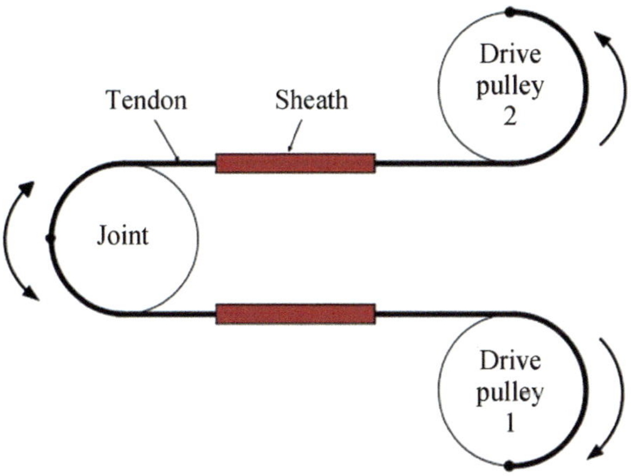

Fig. 2.5 Double tendon-sheath transmission mechanism: independent antagonistic configuration

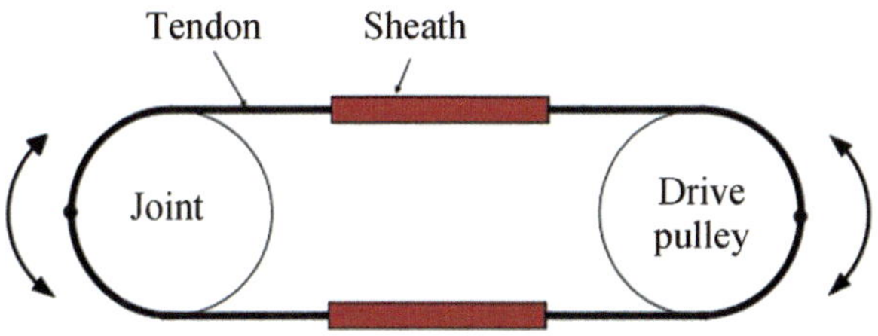

Fig. 2.6 Double tendon-sheath transmission mechanism: coupling drive configuration

in robotic systems operating in confined spaces. For example, the master-slave endoscopic surgical robot MASTER is developed by flexible tendon-sheath transmission mechanism, with a maximum outer diameter of 16 mm [11]. The endoscopic robot has successfully performed endoscopic mucosal dissection in animals. The scorpion endoscopy robot of Kyushu University in Japan also adopted the tendon-sheath transmission mechanism to achieve slender and compliant transmission [12]. However, it's important to note that the interaction between the two tendons during transmission adds complexity to the process. This interaction complicates the transmission of both displacement and torque, posing challenges for precise control.

2.3 Multi-tendon-sheath Transmission System

In robot systems actuated by tendon-sheaths, achieving multi-degree-of-freedom motion in the end effector often requires multiple sets of tendon-sheath mechanisms. These systems, found in applications like minimally invasive surgical robots, robotic arms, and multi-joint exoskeletons, utilize various configurations of single or double tendon-sheath mechanisms to enable multiple joint movements. Despite the lightweight and flexible nature of tendon-sheath mechanisms, the inclusion of cables and sheaths imposes constraints on the outer diameter and flexibility of the slender transmission body. Furthermore, employing multiple sets of driving motors not only increases system cost but also adds complexity, posing challenges for meeting the demands of even slimmer and lighter robot applications.

To fully capitalize on the economic, slender, and flexible advantages of tendon-sheath transmission, while considering the motion characteristics of robot systems, we propose two innovative multi-tendon-sheath transmission systems: the multi-tendon with one sheath transmission system and the multi-tendon-sheath with a single motor transmission system. These innovations aim to achieve a more flexible and lightweight drive for robots.

2.3.1 Multi-tendon-sheath Transmission System with One Sheath

(1) Transmission configuration

In contrast to the traditional tendon-sheath transmission mechanism, the key feature of this multi-tendon-sheath transmission system lies in utilizing just one sheath to transmit two or more cables. These cables are housed in a specially designed multi-channel sheath, eliminating the need for redundant sets of sheaths. As illustrated in Fig. 2.7, the two cables of a double tendon-sheath mechanism can be accommodated in a single sheath to construct the multi-tendon-sheath transmission system. This innovative approach effectively reduces the dimensions, mass, and stiffness of the transmission structure. It is particularly well-suited for miniature or slender robots with small sizes and limited movement spaces.

The sheaths housing multiple cables can directly constitute the slender and flexible body of the robot, leading to a reduction in robot size and an expanded range of applicable spaces. As depicted in Fig. 2.8, a minimally invasive surgical robot constructed with the multi-tendon-sheath structure incorporates multiple drive cables for end-effector joints within a single multi-channel sheath. This results in a robot with a flexible and slender body. The slender design of the minimally invasive surgical robot facilitates the performance of intricate surgical procedures in tighter anatomical channels, offering enhanced flexibility, strong passability, and reduced invasiveness.

To minimize the size and weight of the transmission mechanism and ensure its compliance and transmission efficiency, the arrangement of multiple cables in a sheath is very important. How to select the structure, material, and manufacturing of the sheath and the cable to reduce the friction and interference between the cables under the limited size requires further research, which is seldom studied in ordinary tendon-sheath transmission.

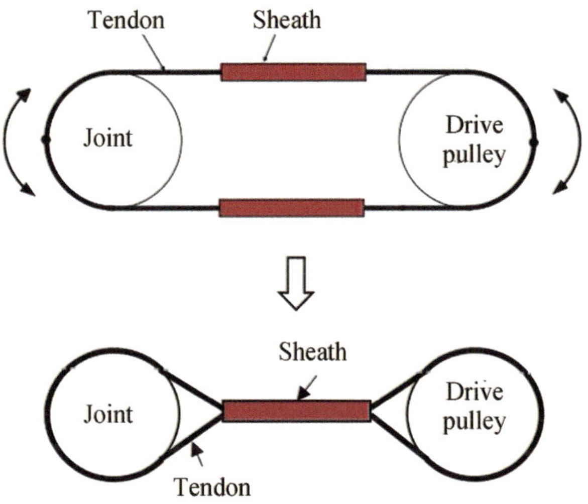

Fig. 2.7 Double tendon-sheath transmission changes to multi-tendon-sheath transmission

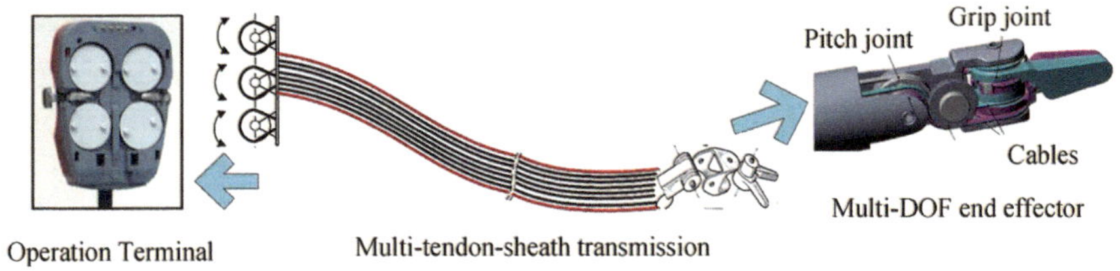

Fig. 2.8 The multi-tendon-sheath transmission system for surgical robot

Constructing a slender multi-tendon-sheath mechanism hinges on the manufacturing method, material selection, and structure design. Typically, lightweight and soft materials with high compressive rigidity, such as plastic, composite rubber, and elastic materials, can serve as sheathing materials. The multi-channel sheathing structure can be created through molding, 3D printing, or laser cutting. Moreover, the mechanical challenges associated with multi-tendon-sheath transmission are intricate. Analyzing bending capacity, stiffness changes, and force transmission are crucial performance evaluation indicators for the multi-tendon-sheath mechanism. This analysis provides design criteria tailored to different load applications.

(2) Variable stiffness mechanism

This multi-tendon-sheath transmission mechanism presents an innovative approach for designing slender robots intended for operation in confined spaces. Examples include surgical robots designed for natural orifices and inspection robots used in narrow industrial pipelines. The multi-tendon-sheath mechanism proves useful in constructing the slender body of these robots. In such applications, the body of the robot must exhibit flexibility to smoothly navigate through curved passages. Once the robot reaches its intended position, the slender body is crucial to providing a stable base for precise manipulation of the end effector. Given the importance of the multi-tendon-sheath transmission in constructing the slender body, variations in stiffness and control become crucial considerations in the structure design of the multi-channel sheath. To address this, a hyperelastic continuum structure can be employed as the bending sheath to achieve specific stiffness design. Introducing asymmetric cuts on the surface of the continuum structure allows for changes in bending stiffness to adapt to different environmental constraints.

To achieve a varied and controllable stiffness of the outer sheath, we proposed a new structure of multi-tendon-sheath transmission with variable stiffness, as shown in Fig. 2.9. The sheath is composed of small ball joints and connection joints in series. The surrounding cables can adjust the bending shape of the slender robot or transmit the end movement. Single or double cables can be set up in the same channel. The tension cable in the center can adjust the bending stiffness of the slender body and improve the stability and precision of terminal positioning operation.

As Fig. 2.10 shows, when all the units are pressed by the tension cable, the bending stiffness of the entire tendon-sheath mechanism can be improved by the friction between the contact interfaces of these joints, and the maximum lateral load can be obtained as:

$$F = \frac{\mu F_N R\,(\sin\theta_1\cos\theta_1 - \sin\theta_2\cos\theta_2)}{\pi l\,(\sin\theta_2 - \sin\theta_1)} \quad (2.3.1)$$

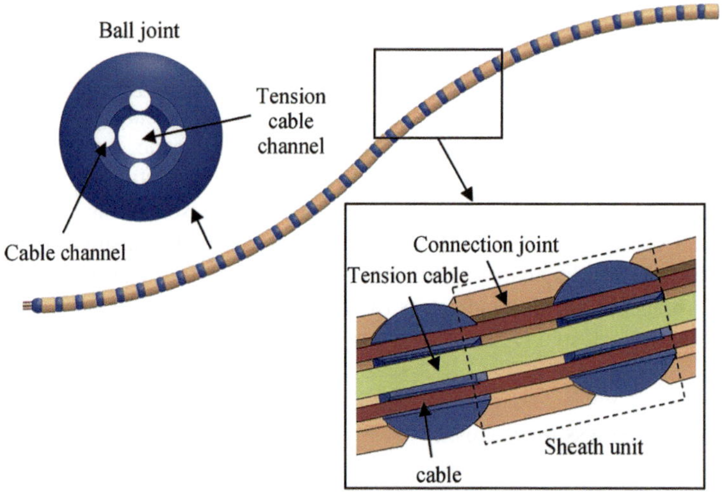

Fig. 2.9 The multi-tendon-sheath transmission mechanism with variable stiffness

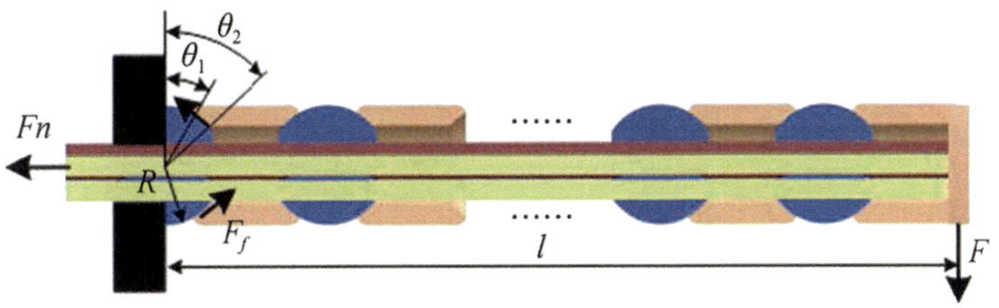

Fig. 2.10 Variable stiffness analysis of the multi-tendon-sheath transmission mechanism

Therefore, by increasing the pre-tightening force of the central tension cable and increasing the friction of the contact interface, the resist lateral loads can be effectively improved. The new structure has the advantages of compact structure, adaptability, and controllable stiffness. The stiffness can be precisely adjusted by precisely controlling the pre-tightening force of the tension cable. However, the interaction between the internal cables and the sheath is complex. To this end, future research will focus on the force/position transfer characteristics and structural stiffness control techniques.

2.3.2 Multi-tendon-sheath Transmission System with One Drive Motor

(1) Transmission configuration

The priciest and heaviest component in the tendon-sheath transmission system is typically the drive unit, commonly a motor-reducer unit. When employing tendon-sheath transmission for driving multi-degree-of-freedom joints, it often necessitates multiple motors to propel a set of single or double tendon-sheath transmission mechanisms, facilitating the intricate movement of these end joints. While the tendon-sheath transmission permits the separate design of the driving device and end joints, ensuring the compactness, flexibility, and essential compliance for applications like surgical robots and search and rescue robots, the challenge lies in reducing the size and weight of the driving mechanism. Consequently, these robots commonly feature a relatively fixed large-size drive console. Consequently, for robots requiring increased flexibility and mobility with multiple degrees of freedom, the use of multiple motors may result in a bulky mechanism with insufficient compliance. For instance, a wearable exoskeleton device often includes a backpack weighing tens of kilograms, serving as storage for motors, reducers, and power supplies.

To address this, we propose a multi-tendon-sheath system with a single motor to drives multiple joints, as shown in Fig. 2.11. In simple, the multiple end joints are driven by the same motor, and each joint has an independent tendon-sheath transmission path, and different elasticities can be configured at the end to achieve different

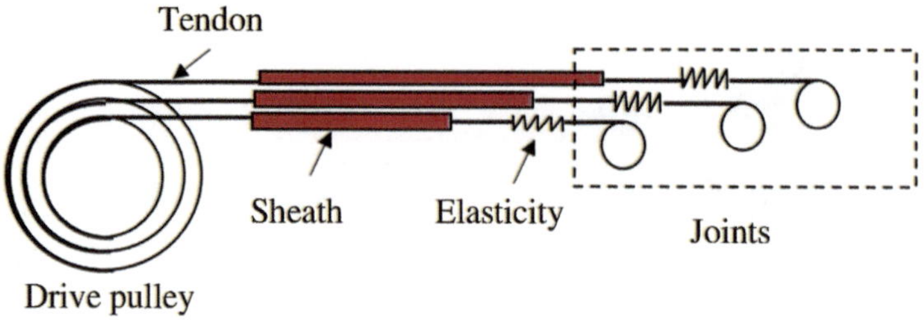

Fig. 2.11 The multi-tendon-sheath transmission mechanism with one drive motor

distributions of force and displacement. To enable multi-degree-of-freedom movement in the end joints, customization options are available within this transmission mechanism:

(1) Each end joint can be fitted with either single tendon-sheath transmission or double tendon-sheath transmission based on specific movement requirements;
(2) The drive pulley linked to each joint can be designed with varying radii or even take the form of cams, connecting rods, etc., particularly suitable for periodic movements;
(3) The elasticity at the end joint can be configured as a nonlinear elastic unit or a variable elastic mechanism, adjusting to the motion requirements.

This type of multi-tendon-sheath system minimizes the need for motors, effectively reducing the robot's weight and making it lighter and more energy-efficient. However, the coupling of each joint motion is high, making it better suited for multi-joint motions with significant correlation, such as the periodic motion of the joints in human lower limbs during walking or the synchronized continuous motion of finger joints.

(2) Variable transmission mechanism

To enhance the flexibility and functional adaptability of multiple joints driven by a single motor, we propose a multi-tendon-sheath transmission system for an upper limb exoskeleton, as depicted in Fig. 2.12. Two tendon-sheath transmission branches simulate the biceps and are driven by a single motor. Similar to the biceps structure, the multi-tendon-sheath system features two origin points and an end point, enabling more natural arm actuation. Each actuation branch is equipped with serial springs of varying stiffness and a micro clutch for switching between actuation and release. This setup achieves coordinated drive for multiple joints.

The clutch plays a crucial role in coordinating the actuation of the multi-tendon-sheath mechanism. It must quickly and effectively clamp or loosen the cable, with the maximum clamping force directly influencing the driving force of each branch.

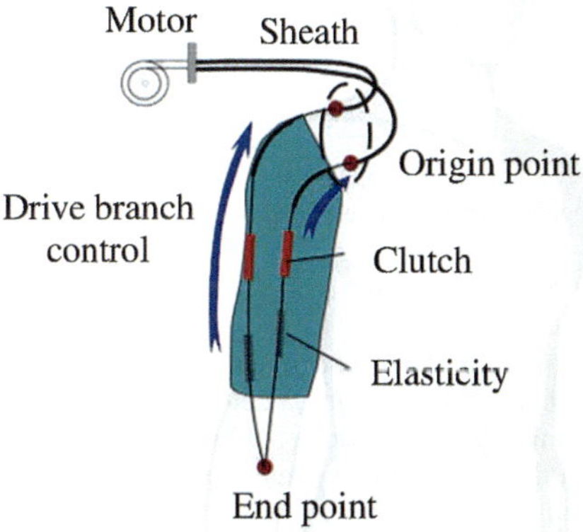

Fig. 2.12 The multi-tendon-sheath transmission mechanism for upper extremity exoskeleton

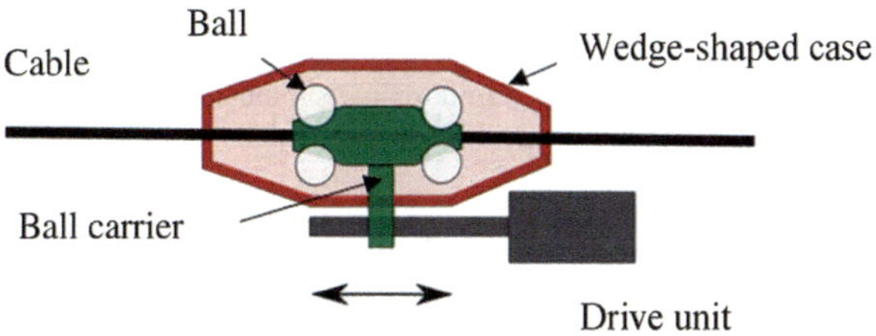

Fig. 2.13 The tristable micro clutch

Thus, the clutch needs to be lightweight and small, ensuring a substantial clamping force for arm actuation. Employing the unidirectional tension property of cables, we design a tristable micro clutch shown in Fig. 2.13. The ball carrier propels the steel ball to the wedge-shaped end, clamping the cable with the wedge-shaped case and enabling self-locking clamping in one direction.

To achieve rapid two-way clutching, a mechanism with double-ended balls is devised. The ball carrier is driven left and right in the wedge-shaped housing via a motor or electromagnet, thereby locking or releasing the cable in both directions. When the ball is in the middle of the wedge-shaped case, the cable can move freely in two directions. Additionally, the response time of cable clamping and loosening, along with the maximum clamping force, directly influences the speed and effectiveness of joint motion coordination control. Therefore, considerations for optimizing the clutch mechanism include the force-position characteristics of the motor/electromagnetic drive, the size and weight of the ball carrier, and the wedge shape of the case.

References

1. Lin C, Xingsong W, Fuxiang T (2009) Tendon-sheath actuated robots and transmission system. In: 2009 international conference on mechatronics and automation. IEEE, pp 3173–3178 (2009)
2. Palli G, Melchiorri C (2006) Model and control of tendon-sheath transmission systems. In: Proceedings 2006 IEEE international conference on robotics and automation. ICRA 2006. IEEE, pp 988–993
3. Sampedro C, Rodriguez-Ramos A, Bavle H et al (2019) A fully-autonomous aerial robot for search and rescue applications in indoor environments using learning-based techniquesd. J Intell Robot Syst 95:601–627
4. Xiao X, Murphy R (2018) A review on snake robot testbeds in granular and restricted maneuverability spaces. Robot Auton Syst 110:160–172
5. Yin M, Shang D, Xu T et al (2021) Joint modeling and closed-loop control of a robotic hand driven by the tendon-sheath. IEEE Robot Autom Lett 6(4):7333–7340
6. Kang B, Choi H, Lee H et al (2019) Exo-glove poly II: a polymer-based soft wearable robot for the hand with a tendon-driven actuation system. Soft Robot 6(2):214–227
7. Hofmann U, Bützer T, Lambercy O et al (2018) Design and evaluation of a bowden-cable-based remote actuation system for wearable robotics. IEEE Robot Autom Lett 3(3):2101–2108
8. Wu Q, Chen B, Wu H (2019) Neural-network-enhanced torque estimation control of a soft wearable exoskeleton for elbow assistance. Mechatronics 63:102279
9. Ding Y, Kim M, Kuindersma S et al (2018) Human-in-the-loop optimization of hip assistance with a soft exosuit during walking. Sci Robot 3(15):eaar5438
10. Wu Q, Wang X, Chen L et al (2014) Transmission model and compensation control of double-tendon-sheath actuation system. IEEE Trans Ind Electron 62(3):1599–1609
11. Ho K, Phee S, Shabbir A et al (2010) Endoscopic submucosal dissection of gastric lesions by using a master and slave transluminal endoscopic robot (MASTER). Gastrointest Endosc 72(3): 593–599
12. Kalloo A, Singh V, Jagannath S et al (2004) Flexible transgastric peritoneoscopy: a novel approach to diagnostic and therapeutic interventions in the peritoneal cavity. Gastrointest Endosc 60(1):114–117

Chapter 3
Transmission Characteristics of Tendon-Sheath System

Abstract As the tendon-sheath system is compliant and has low stiffness, control of these tendon-sheath systems could be challenging. On one hand, they have a lot of nonlinear factors including the compliance and friction between the tendon and sheath; on the other hand, due to spatial restrictions and disinfection requirements, these systems usually lack corresponding sensory feedback. These challenges limit the application of tendon-sheath systems in situations requiring precise control. To solve the problems, many studies on modeling and analysis of tendon-sheath transmission characteristics are reported. This chapter introduces the transmission characteristics of both single tendon-sheath and double tendon-sheath mechanisms, providing a foundation for modeling other tendon-sheath systems. To aid in the design and optimization of tendon-sheath systems, the chapter analyzes the main factors influencing tendon-sheath transmission. Furthermore, the most important factor, the bending curvature of sheath is discussed and a curvature optimization method is introduced.

3.1 Transmission Characteristics of Single Tendon-Sheath Mechanism

The tendon-sheath mechanism has a very low stiffness, and the friction between the tendon and the sheath is complex, such that it has serious nonlinear transmission characteristics. Many scholars have carried out research on the transmission characteristics of the tendon-sheath mechanism [1, 2].

Figure 3.1 shows the friction in the tendon-sheath mechanism during the change of the movement direction, in which the solid arrows represent the dynamic friction and the hollow arrows represent the static friction [3]. As the cable is consistently tensioned during movement, the tendon slides on the convex side of the inner wall of the sheath. In transmission process A, as the input force gradually increases and the entire tendon moves, the direction of friction opposes the movement direction of the tendon; When the input force gradually decreasing, the movement of the tendon will stop, and at the same time, the dynamic friction in the tendon-sheath mechanism will

X. Wang and Q. Zhang, *Tendon-Sheath Transmitted Light and Slender Robots*,
https://doi.org/10.1007/978-981-96-5581-6_3

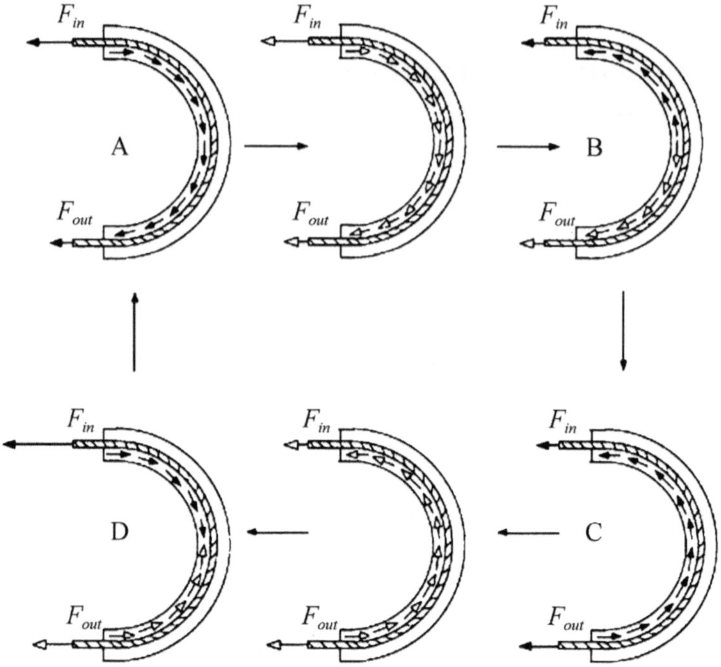

Fig. 3.1 Friction between tendon and sheath during turning [3]. The solid arrows represent dynamic friction, hollow arrows represent static friction

be transformed into static friction, and the whole tendon achieve a balance between the drive force and the static friction; As the input force continues to decrease to process B, the equilibrium state of the tendon part near the input end will be broken first and the tendon begins to move in the opposite direction, while the tendon part near the output end is still in a static state, so the tendon part around the input end has dynamic friction and the part from the output end still has static friction; When the tension of tendon continues to decrease, the whole tendon starts to move in the opposite direction and enters the process C. At this time, the frictions in the tendon-sheath mechanism are all dynamic frictions in the reverse direction. Up to now, the tendon-sheath mechanism has completed the entire reverse movement process, that is, the process A to the process C. Similarly, starting from state C, when the input tension begins to increase, the moving tendon is blocked and eventually comes to stop. At this time, dynamic friction transitions into static friction. As the tension in the input persists, the section of the rope near the driving end undergoes a reversal in motion, leading to the entire tendon executing a reverse motion and completing the movement in the opposite direction. The changing law of friction is just as the process A to process C, while the friction direction is opposite.

To measure the force and displacement transmission characteristics of the tendon-sheath system, an experimental platform was built as Fig. 3.2 shows. A sinusoidal

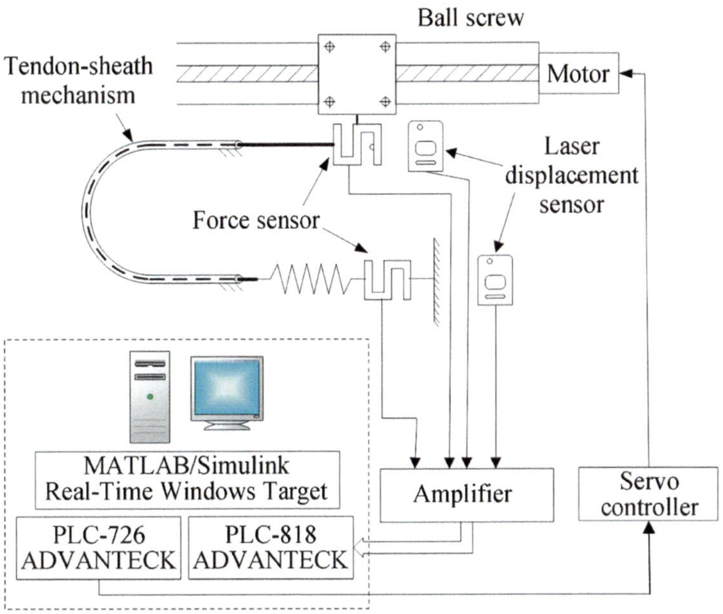

Fig. 3.2 Measurement setup for charactering a tendon-sheath mechanism [10]

periodic excitation signal with low frequency is given to the input motor, such that the motor at the input end can drive the tendon to move slowly and acquire an approximate quasi-static motion state. The static force transmission characteristics of the tendon-sheath system measured under this excitation are shown in Figs. 3.3 and 3.4. Figure 3.3 depicts the friction corresponding to the unit load, while Fig. 3.4 shows the relationship between input and output force. From Fig. 3.3, we can find that the friction also shows periodic changes with the periodic transmission motion. The friction in the tendon-sheath mechanism shows minor changes during the transmission in the same direction, while the friction direction changes with the transmission direction.

Fig. 3.3 The static friction measurement of the tendon-sheath mechanism [10]

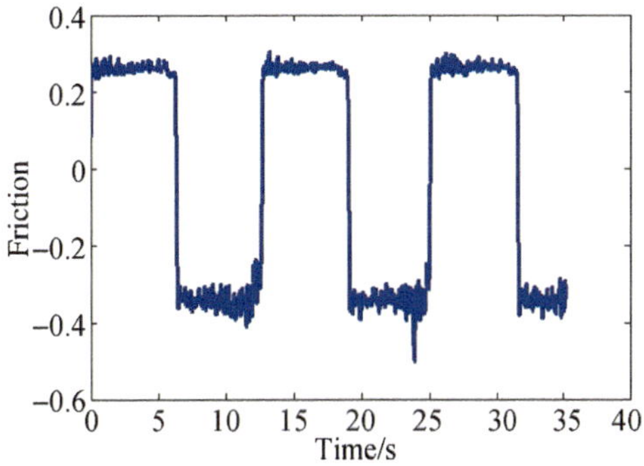

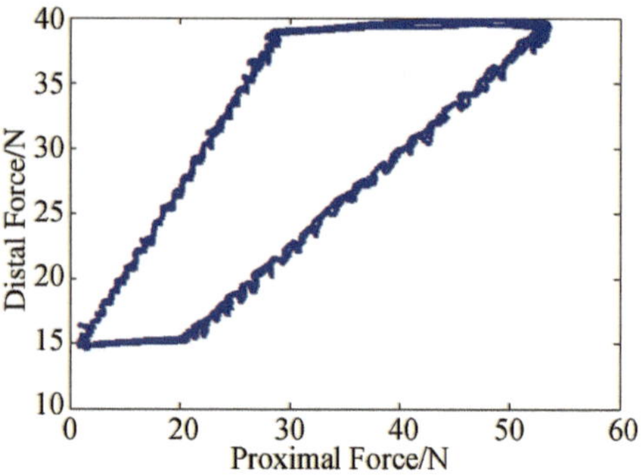

Fig. 3.4 The static input force and output force [10]

However, the magnitude of the friction in the forward and reverse directions is not the same. This is also evident in Fig. 3.4, the interval of the upper and lower approximately horizontal curves reflect the input force for overcoming the friction in the tendon-sheath mechanism, and obviously the two intervals are not the same. Such that the friction leads to a hysteretic characteristic of the force transmission.

The dynamic force transmission characteristics of the tendon-sheath system were tested using the excitation signal with high frequency, such that the motor will pull the tendon to move quickly, making the tendon in a dynamic moving process. During this process, the measured frictions are shown in Figs. 3.5, while 3.6 shows the relationship between the input and output force. Note that there are a number of cycles of loading in Fig. 3.6.

Compared with the static force transmission characteristics, the dynamic force transmission characteristics are gradual and smooth when the movement direction of the tendon changes, and the transmission appeared the stribeck phenomenon.

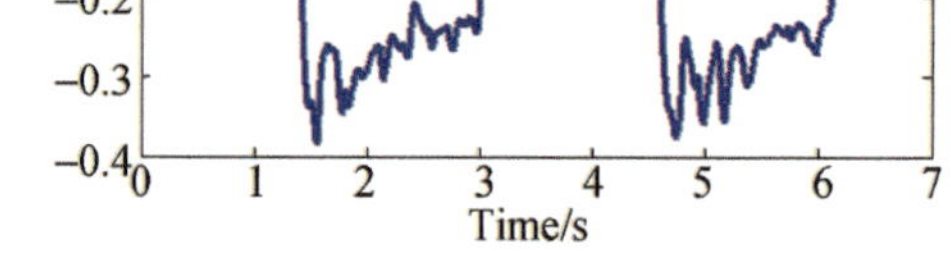

Fig. 3.5 The dynamic friction measurement of the tendon-sheath mechanism [10]

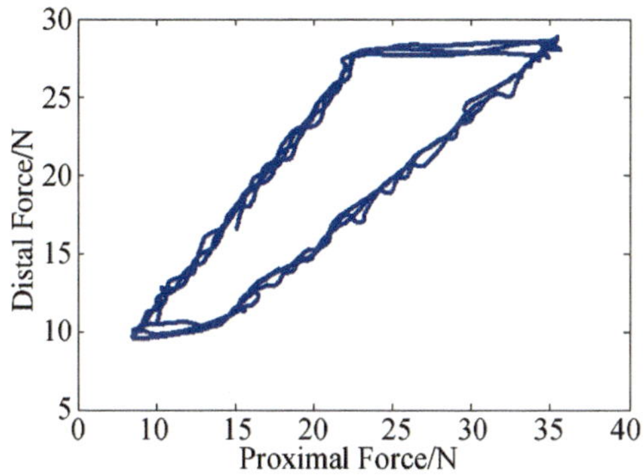

Fig. 3.6 The dynamic input force and output force [10]

3.2 Transmission Characteristics of Double Tendon-Sheath Mechanism

According to the previous analysis results [4], the system parameters in Table 3.1 were employed for simulation. The simulated transfer characteristics of displacement and torque are shown in Figs. 3.7 and 3.8 respectively. Notably, the simulation results exhibit three distinct stages. Stages 3 and 6 represent active working phases, where the input and output are proportional. Conversely, Stages 4 and 7 indicate delay stages, where the output remains constant despite changes in the input. The width of these delay stages decreases with an increase in the position of the pulley at the drive end during a change in direction. Stages 5 and 8 are transitional phases, revealing an approximately linear relationship between input and output, with the slope varying based on the width of the delay stage. A larger width corresponds to a smaller slope. The larger the width, the smaller the slope. In Figs. 3.7 and 3.8, Stage 1 captures the state where the drive pulley moves while the end joint remains stationary under preload conditions. Stage 2 illustrates a scenario where one cable overcomes the preload to drive the end joint, while the other cable has not yet surpassed the applied preload.

Figure 3.9 shows the change relationship of force transmission at both ends of two cables. Notably, the transmission characteristics between the two ends of the cables differ when they are coupled and driven compared to a single cable. The

Table 3.1 Parameter value of system simulation

Preload (N)	10	End load scale factor (Nm/rad)	5
Sheath bending radius (m)	0.2	Cable diameter (mm)	0.8
Sheath bending angle (rad)	2π	Young's modulus (Gpa)	140
Pulley radius (mm)	15	Friction coefficient	0.1

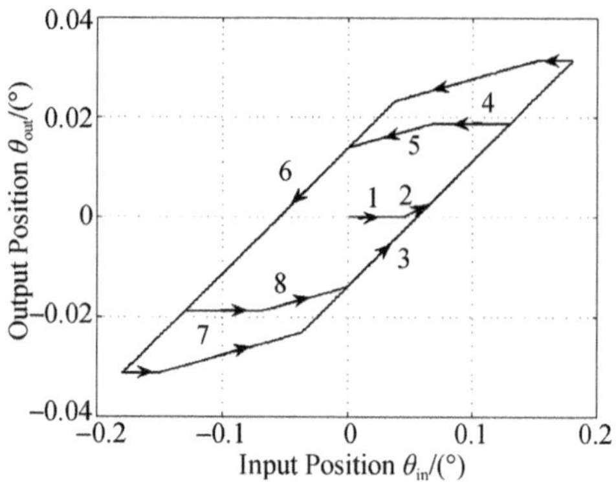

Fig. 3.7 Simulation results of displacement transmission characteristics of double tendon-sheath coupling transmission [3]

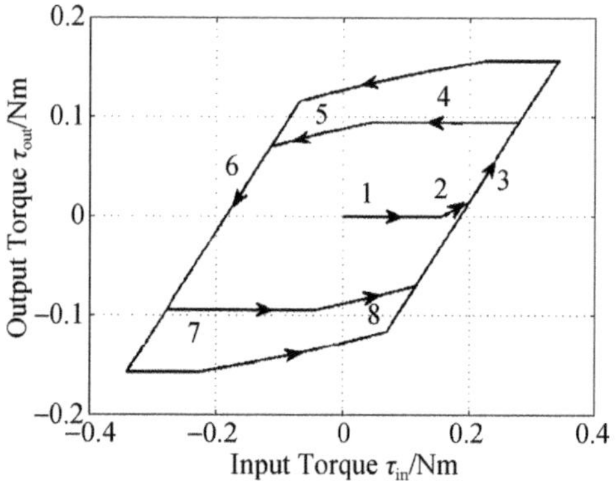

Fig. 3.8 Simulation results of torque transmission characteristics of double tendon-sheath coupling transmission [3]

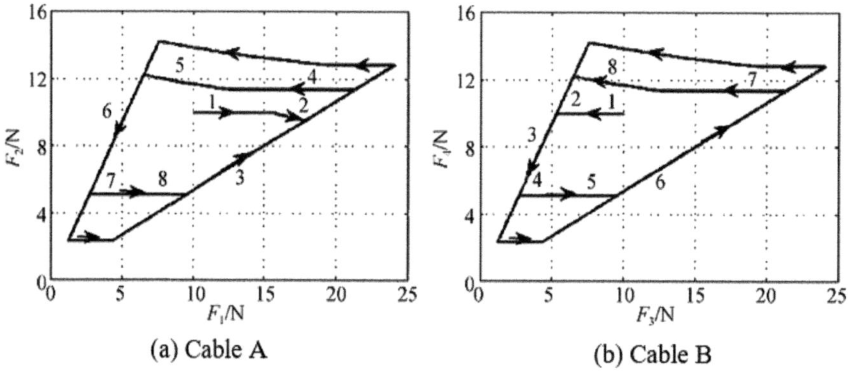

Fig. 3.9 Simulation results of force transmission characteristics of double tendon-sheath coupling transmission [3]

output force experiences variation in the middle of the stretching and relaxation phases, with a distinct "peak" stage characterized by a sudden force increase. This peak stage emerges during the transition phase, driven by the movement of the input pulley, influencing the force in both cables. However, the force alteration in one cable reaches the end pulley first, initiating its movement, and subsequently pulling the other cable. This dynamic, where one cable pulls another, is absent in single cable transmission. The width of the peak stage decreases as the force value diminishes during the initial stages of force change.

3.3 Influencing Factors of Transmission Characteristics

(1) Change of movement speed

In this group of experiments, the bending angle of the sheath is π and the bending radius is 98 mm. The control motor drives the steering wheel to move at the speed of 5 degrees, 10 degrees and 15 degrees per second respectively. The transmission characteristics of the system are shown in Fig. 3.10. It can be seen that the displacement and torque transmission characteristics of the system hardly change when the movement speed of the steering wheel changes.

(2) Change of sheath's bending radius

In this group of experiments, the bending angle of the sheath is π, and the motion speed of the rudder wheel is 5 degrees per second. Change the bending radius of the sheath to 98 mm, 72 mm and 50 mm respectively. The measured transmission

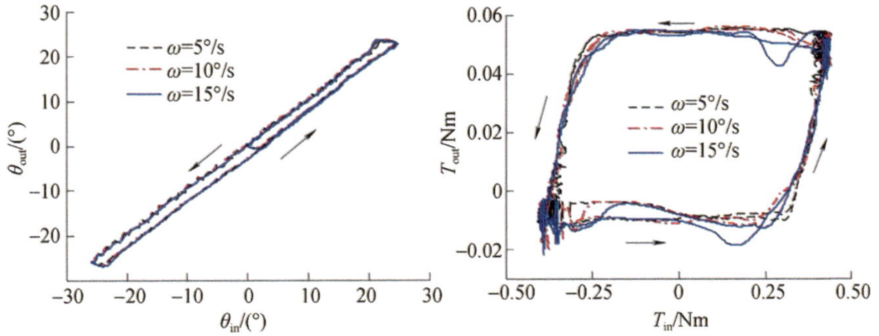

Fig. 3.10 Transmission characteristics of double tendon-sheath mechanism when the motion speed changes [3]

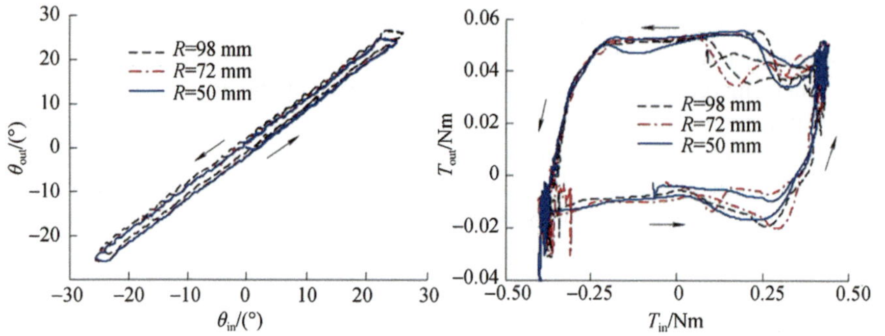

Fig. 3.11 Transmission characteristics of double tendon-sheath mechanism when the bending radius of sheath changes [3]

characteristics of the system are shown in Fig. 3.11. It can be seen that the displacement and torque transmission characteristics of the system hardly change when the bending radius of the sheath changes.

(3) Change of sheath bending angle

In this set of experiments, the sheath's bending radius is 98 mm, and the movement speed is 10 degrees per second. The sheath is bent into $\pi/2$, π and $3\pi/2$ respectively. The measured transmission characteristics of the system are shown in Fig. 3.12. The displacement transmission characteristics are shown in Fig. 3.12, reveal that the width of displacement characteristics gradually increases with the sheath's bending angle, although the change is not pronounced. This is primarily attributed to the cable diameter of 0.8 mm, which results in a relatively constant tensile force of about 20N. Consequently, the elongation change in the cable remains relatively small.

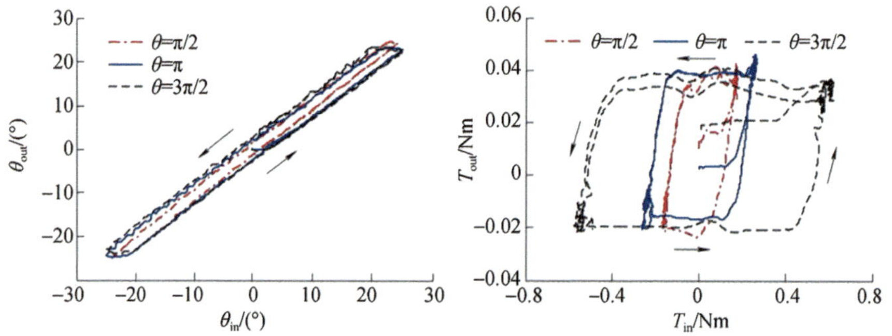

Fig. 3.12 Transmission characteristics of double tendon-sheath mechanism when sheath bending angle changes [3]

The characteristics of torque transmission are shown in Fig. 3.12. It can be seen that with the increase in casing bending angle, under the same output torque, a larger sheath bending angle necessitates greater torque input. This results in a wider "gap" in the torque transmission characteristics. This underscores the importance of minimizing the casing's bending angle in the design of the double tendon-sheath coupling transmission system to mitigate torque loss.

(4) Change of preload

In this set of experiments, the cable's pretension is adjusted by modifying the depth of the pretension-adjusting part screwed into the tension sensor. However, ensuring complete consistency of tension along the sheath direction proves challenging. The sheath's bending angle and radius are both 98 mm, while the movement speed of the steering wheel is set at 10 degrees per second. The measured transmission characteristics of the system are shown in Fig. 3.13. The displacement transfer characteristics shown in Fig. 3.13 reveal that the gap in these characteristics gradually diminishes with increasing preload. However, as shown in Fig. 3.13, the increase of preload results in an expanded "clearance" in torque transmission characteristics. This highlights the necessity, during system adjustments post-design, to strike a balance between the precision of displacement transmission and the mitigation of torque transmission loss. This aims to minimize transmission clearance while ensuring the input torque remains within reasonable limits.

(5) Change of lubrication

In this set of experiments, the bending angle of the sheath is π, the bending radius is 98 mm, and the motion speed of the steering wheel is 15 degrees per second. In one set of experiments, dry friction was introduced between the cable and the sheath,

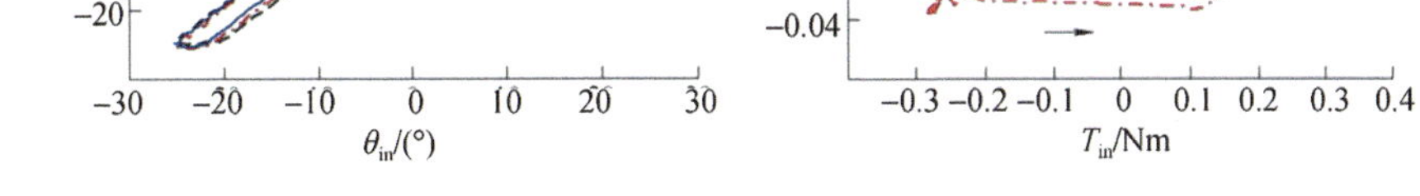

Fig. 3.13 Transmission characteristics of double tendon-sheath mechanism when the preload of cable changes [3]

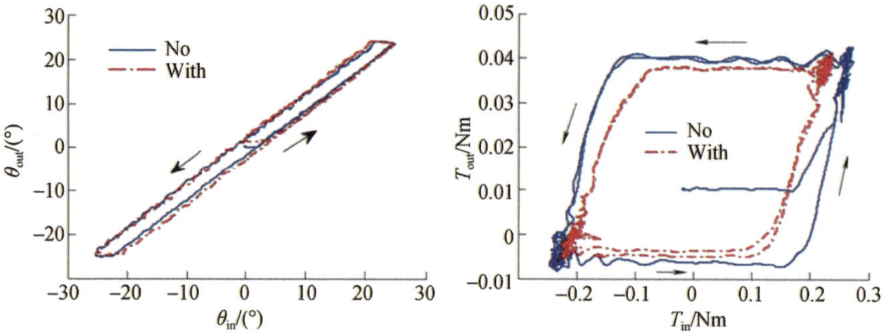

Fig. 3.14 Transmission characteristics of double tendon-sheath mechanism under lubrication changes [3]

while in the other group, grease lubrication (Monroe advanced bearing professional grease) was applied in the tendon-sheath mechanism. From the position transmission characteristics shown in Fig. 3.14, it can be seen that there is no apparent alteration in the characteristics following the introduction of grease lubrication between the cable and the sheath. However, the torque transfer characteristics in Fig. 3.14 reveals a notable reduction in the width of the torque transfer transition region after the application of grease lubrication. Under identical output torque, the system requires significantly less input torque post-lubrication, indicating an enhanced torque output efficiency. This shows that adding lubrication in the tendon-sheath transmission system can effectively improve the torque output efficiency.

3.4 Comprehensive Optimization of Transmission Distance and Space Angle

The cumulative bending angle significantly influences the transmission characteristics of the tendon-sheath mechanism. An increase in the bending angle results in heightened friction and diminished transmission efficiency [5, 6]. Therefore, arrangement of the tendon-sheath transmission path is crucial for complex multi-degree-of-freedom motion systems. To enhance transmission efficiency and mitigate the impact of bending angle changes during movement, the tendon-sheath transmission path must be optimally organized. To explore this, we focus on the multi-joint lower extremity exoskeleton as a representative multi-degree-of-freedom motion system for research on optimizing the tendon-sheath space transfer path [7, 8].

3.4.1 Full Curvature of Tendon-Sheath System

The arbitrary curve in space, which can be seen as the trajectory of a particle in space corresponding to a change in time t[9], whose parametric equations are as follows.

$$\begin{cases} x = x(t) \\ y = y(t) \quad \mathrm{a} < t < b \\ z = z(t) \end{cases} \tag{3.4.1}$$

The above equation is expressed in vector form as

$$r(t) = (x(t), y(t), z(t)) \tag{3.4.2}$$

Assume that for any $t \in (a, b)$, $r'(t) \neq 0$. When the arc length is used as the curve parameter. The arc length function of the curve in Eq. (3.4.2) is expressed as

$$s(t) = \int_{t_0}^{t} \left|r'(t)\right| dt = \int_{t_0}^{t} \sqrt{\dot{x}(t)^2 + \dot{y}(t)^2 + \dot{z}(t)^2} dt \tag{3.4.3}$$

Combining Eqs. (3.4.2) and (3.4.3), we can get the formula for this curve about the arc length.

$$r(s) = (x(s), y(s), z(s)) \tag{3.4.4}$$

The unit tangent vector at any point P on the curve is denoted as $r'(s)$, and the second-order derivative of any point on its curve $r''(s)$ is expressed as the curvature of the space curve at point P.

$$\kappa(s) = \left|r''(s)\right| \tag{3.4.5}$$

Define the full curvature of the space curve as

$$\theta = \int_{0}^{L} \kappa(s) ds \tag{3.4.6}$$

where L is expressed as the arc length of the curve. The full curvature reflects the total curvature of the curve.

To optimize the tendon-sheath transfer path of the lower limb rehabilitation exoskeleton in a more effective way, we simplify it as the optimization of plane curves. This reduction allows us to treat arbitrary curves in the plane as parametric equations varying with time t.

$$\begin{cases} x = x(t) \\ y = y(t) \end{cases} \quad \text{a} < t < b \tag{3.4.7}$$

Similarly, assuming that the plane curve is smooth and quadratic differentiability, we can get the arc length function of the plane curve as

$$s(t) = \int_{t_0}^{t} \sqrt{\dot{x}(t)^2 + \dot{y}(t)^2} dt \tag{3.4.8}$$

Let $\alpha(t)$ be the tangent inclination at the point $P(x(t), y(t))$ on the curve, and $\Delta\alpha = \alpha(t + \Delta t) - \alpha(t)$ denotes the increment of tangent inclination when the point P on the plane curve is moved to the point $Q(x(t + \Delta t), y(t + \Delta t))$. The arc length $\hat{PQ}$ is expressed as Δs, then the average curvature of the arc length $\hat{PQ}$ is expressed as follows

$$\bar{\kappa} = \left| \frac{\Delta\alpha}{\Delta s} \right| \tag{3.4.9}$$

$\bar{\kappa}$ is the average curvature of the curve at P,

$$k = \left| \lim_{\Delta t \to 0} \frac{\Delta\alpha}{\Delta s} \right| = \left| \lim_{\Delta s \to 0} \frac{\Delta\alpha}{\Delta s} \right| = \left| \frac{d\alpha}{ds} \right| \tag{3.4.10}$$

Assuming that the curve is always smooth, we have

$$\alpha(t) = \arctan \frac{y'(t)}{x'(t)} \tag{3.4.11}$$

As $x(t)$ and $y(t)$ are second-order differentiable, and the arc differential $ds = \sqrt{(dx)^2 + (dy)^2}$, which can be derived as

$$\frac{d\alpha}{ds} = \frac{\alpha'(t)}{s'(t)} = \frac{x'(t)y''(t) - x''(t)y'(t)}{\left[x'^2(t) + y'^2(t)\right]^{\frac{3}{2}}} \tag{3.4.12}$$

The curvature expression is

$$\kappa = \frac{|x'y'' - x''y'|}{\left(x'^2 + y'^2\right)^{\frac{3}{2}}} \tag{3.4.13}$$

Similarly, the full curvature of the plane curve is defined as

$$\theta = \int_0^L \kappa(s) ds = \int_a^b \frac{|x'y'' - x''y'|}{\left(x'^2 + y'^2\right)^{\frac{3}{2}}} \cdot \sqrt{x'^2 + y'^2} dt = \int_a^b \frac{|x'y'' - x''y'|}{\left(x'^2 + y'^2\right)} dt \tag{3.4.14}$$

3.4.2 Exoskeleton Joint Transmission Path Optimization

(1) Hip joint path optimization

In the lower limb exoskeleton mechanism, the distance from the hip joint to the waist-hip motor module remains constant throughout the entire gait walking process. The optimal bending path is determined by the angle between the output end of the hip motor and the execution end of the hip joint, the specific theoretical analysis is as follows.

As shown in Fig. 3.15, the xoy coordinate system is established with the center of the hip joint, and the hip joint sheath support points are designed as the figure shows, with considering the structural design and the spatial interference. Point $A(x_1, y_1)$ is the right end sheath support fixation point at the hip joint execution, A' is the left end sheath support fixation point, A is the right end sheath support fixation point, point D is the midpoint of $A'A$, and point $O_0(x_2, y_2)$ is the center point of the drive disc at the output end of the hip motor, thus the coordinates of two points A and O_0 are known, and the slope of the sheath at point A is 0, that is, $f'(x_1, y_1) = 0$. r_1 is the radius of the drive disc, and r_2 is the drive radius where the sheath fixed. Then make a tangent line to the circle r_1 from point A and intersect at point B. The line AB intersects the circle corresponding to r_2 at point C. The slope of the sheath CD at point C is $\tan\theta_1 = \tan - (90° - \angle CAE)$, and the length of the segment AO_0 is

$$l_{AO_0} = \sqrt{(x_2 - x_1)^2 + (y_2 - y_1)^2} \tag{3.4.15}$$

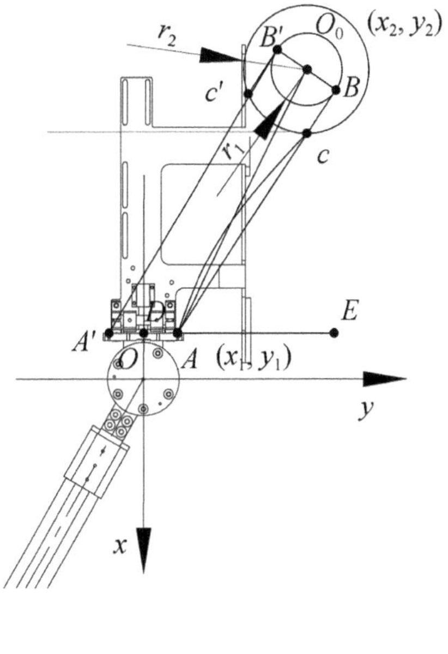

Fig. 3.15 Hip joint tendon-sheath path optimization analysis diagram [11]

In addition,

$$\angle O_0AE = \arctan\left(\frac{y_2 - y_1}{x_1 - x_2}\right) \tag{3.4.16}$$

$$\angle O_0AC = \arcsin\frac{r_1}{l_{AO_0}} \tag{3.4.17}$$

Combining Eqs. (3.4.15), (3.4.16) and (3.4.17), we have

$$\angle CAE = \angle O_0AE - \angle O_0AC \tag{3.4.18}$$

According to these analysis we know that: the angle of the forward tangent of the smooth curve at O is θ_0, and $\int_a^b \text{kds} \geq \theta_0$. The positive angle of the sheath curve passing through points C and A is

$$\theta_2 = \frac{\pi}{2} - \frac{\pi}{180} \times \angle CAE \tag{3.4.19}$$

Therefore, the full curvature of the curve CA having a slope of $\tan\theta_1 = \tan - (90° - \angle CAE)$ at point C, and the full curvature of the slope of 0 at point A is

$$\theta_h = \int_0^L \kappa(s)ds \geq \frac{\pi}{2} - \frac{\pi}{180} \times \angle CAE \tag{3.4.20}$$

Such that the full curvature is at least θ_2, and its length slightly exceeds the distance of CA. The preload mechanism at point C can change the straight line distance between the two endpoints of the sheath, whten it becomes shorter, the full curvature changes accordingly. From the Eq. (3.4.20), we understand that as the y-axis distance from point C to point A decreases, the x-axis distance increases, and the full curvature of the casing between AB diminishes. Therefore, in the structural design of the motor end preload mechanism, it is advisable to minimize the impact on the upper end of the waist while avoiding interference with other structures.

B' for BO_0 and r_1 corresponds to the intersection of the circle, and $C''B'$ for r_1 corresponded to the tangent line at point B, we can calculate the full curvature of the sheath at $C''A'$ in a similar way. For this situation, if the difference between the diameters of the drive disc and the actuator disc is small, we typically choose two points, B and B', parallel to the circle tangent to DO_0 and r_1. This selection accounts for the minimal difference in length between line segment $A'A$ and the circle diameter corresponding to r_1. The method for calculating full curvature remains the same as described earlier.

(2) Knee joint path optimization

In the lower limb exoskeleton mechanism, the constant distance from the hip axis to the hip motor simplifies the optimization of the hip tendon-sheath path. However, the changing distance from the knee axis to the knee motor poses a challenge. Theoretical analysis, combined with the study of the human motion envelope curve, reveals that the distance from the knee axis to the hip axis remains constant. Therefore, the knee tendon-sheath path is guided through the guide block at the hip joint, following the human motion envelope curve.

Using the axis of the hip joint as the coordinate origin, we establish the xoy coordinate system in Fig. 3.16, point O_1 is the center of the drive disc of the knee motor, and BB' are two tangent points parallel to OO_1. Point $A_1(x_1, y_1)$ is the fixation point on the right side of the knee motor output, and $f_{A_1}{}'(x_1, y_1) = \frac{y_0}{x_0}$, point $B_1(x_2, y_2)$ is the center of the hip joint and coincides with the origin of the coordinate system; point $C_1(x_3, y_3)$ is the right fixation point at the knee. Set the hip flexion angle (forward movement of the thigh) as negative and the hip extension angle (backward movement of the thigh) as positive, thus $f_{C_1}{}'(x_3, y_3) = \tan(-\theta_1)$. Therefore, the problem can be simplified using known conditions, to solve the four times Newton-type Hermite interpolation polynomial through points A_1, B_1, C_1 and the two-point derivatives of A_1 and C_1. The interpolation quotients are shown in Table 3.2.

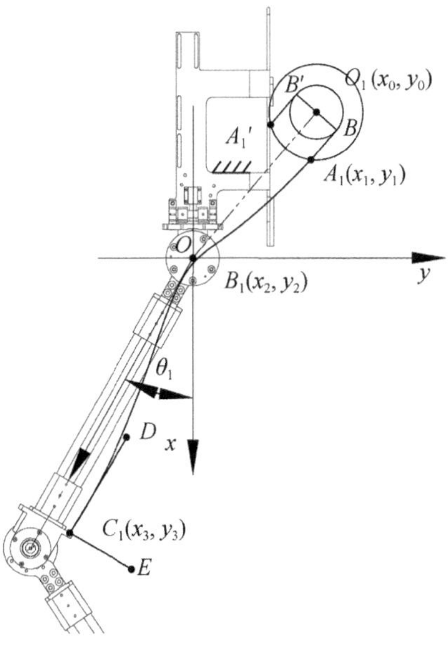

Fig. 3.16 Knee joint tendon-sheath path optimization analysis diagram [11]

Table 3.2 Knee tendon-sheath curve difference quotient table

x_1	$f[x_1]$	$f[x_1, x_1]$	$f[x_1, x_1, x_2]$	$f[x_1, x_1, x_2, x_3]$	$f[x_1, x_1, x_2, x_3, x_3]$
x_1	$f[x_1]$	$f[x_1, x_2]$	$f[x_1, x_2, x_3]$	$f[x_1, x_2, x_3, x_3]$	
x_2	$f[x_2]$	$f[x_2, x_3]$	$f[x_2, x_3, x_3]$		
x_3	$f[x_3]$	$f[x_3, x_3]$			
x_3	$f[x_3]$				

The equations in the table show as Eq. (3.4.21).

$$
\begin{aligned}
& f[x_1] = y_1,\ f[x_2] = y_2,\ f[x_3] = y_3, \\
& f[x_1, x_1] = f'(x_1),\ f[x_1, x_2] = \frac{f[x_2] - f[x_1]}{x_2 - x_1}, \\
& f[x_2, x_3] = \frac{f[x_3] - f[x_2]}{x_3 - x_2},\ f[x_3, x_3] = f'(x_3) \\
& f[x_1, x_1, x_2] = \frac{f[x_1, x_2] - f[x_1, x_1]}{x_2 - x_1} \\
& f[x_1, x_2, x_3] = \frac{f[x_2, x_3] - f[x_1, x_2]}{x_3 - x_1}, \\
& f[x_2, x_3, x_3] = \frac{f[x_3, x_3] - f[x_2, x_3]}{x_3 - x_2}, \\
& f[x_1, x_1, x_2, x_3] = \frac{f[x_1, x_2, x_3] - f[x_1, x_1, x_2]}{x_3 - x_1}, \\
& f[x_1, x_2, x_3, x_3] = \frac{f[x_2, x_3, x_3] - f[x_1, x_2, x_3]}{x_3 - x_1}, \\
& f[x_1, x_1, x_2, x_3, x_3] = \frac{f[x_1, x_2, x_3, x_3] - f[x_1, x_1, x_2, x_3]}{x_3 - x_1}
\end{aligned}
\tag{3.4.21}
$$

From Eq. (3.4.18), the equation of the curve past the points A_1, B_1 and C_1 can be derived as

$$
\begin{aligned}
& H_4(x) = f[x_1] + f[x_1, x_1](x - x_1) + f[x_1, x_1, x_2](x - x_1)^2 \\
& + f[x_1, x_1, x_2, x_3](x - x_1)^2 (x - x_2) \\
& + f[x_1, x_1, x_2, x_3, x_3](x - x_1)^2 (x - x_2)(x - x_3)
\end{aligned}
\tag{3.4.22}
$$

The function $H_4(x)$ derived from Eqs. (3.4.22) with (3.4.8) can be used to determine the theoretical length of the unilateral approximate in-plane trocar for a specific motion stance of the desired knee joint. Since the C_1 position changes with the hip joint, the distance from C_1 to the hip joint is always constant, and the hip joint flexion/extension angle range is small during normal gait walking, the maximum value of $H_4(x)$ obtained at the maximum hip joint flexion/extension angle is the design length of the sheath.

The function $H_4(x)$ derived from Eqs. (3.4.22) and (3.4.14) yields the full curvature of the unilateral approximate in-plane trocar in a given motion posture of the desired knee joint.

(3) Ankle joint path optimization

The lower limb rehabilitation exoskeleton includes passive degrees of freedom in the sagittal plane of the ankle joint. If we introduce active ankle joint motion, driven by a tendon-sheath mechanism with the motor located at the waist, the system becomes more complex. Figure 3.17 shows a coordinate system with the hip joint at its center, in which O_2 is the center of the drive disc of the ankle joint, B and B' are two tangent points parallel to OO_2 and tangent to the disc at the knee joint motor. $A_3(x_1, y_1)$ is the fixed point of the left sheath support at the knee joint motor and its location is known, along with the derivative of the sheath output at A_3. $B_3(x_2, y_2)$ is the center of the hip joint and coincides with the origin of the coordinate system. $C_3(x_3, y_3)$

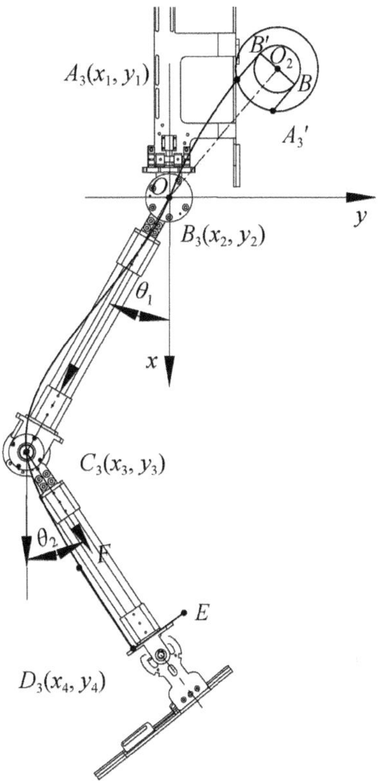

Fig. 3.17 Ankle joint tendon-sheath path optimization analysis diagram [11]

is the center of the knee joint and the $D_3(x_4, y_4)$ is the fixed point of the left sheath support at the ankle joint execution. The knee flexion angle (calf backward motion) is set to positive, thus $f'_{D_3}(x_4, y_4) = \tan\theta_2$. The goal is to optimize this path by finding a 5th Newton-type Hermite interpolation polynomial, $H_5(x)$, based on known A_3, B_3, C_3, D_3 and the derivatives of A_3 and D_3.

Similarly to the knee tendon-sheath path optimization, the curve equation is introduced as

$$\begin{aligned} H_5(x) &= f[x_1] + f[x_1, x_1](x - x_1) + f[x_1, x_1, x_2](x - x_1)^2 \\ &+ f[x_1, x_1, x_2, x_3](x - x_1)^2(x - x_2) \\ &+ f[x_1, x_1, x_2, x_3, x_4](x - x_1)^2(x - x_2)(x - x_3) \\ &+ f[x_1, x_1, x_2, x_3, x_4, x_4](x - x_1)^2(x - x_2)(x - x_3)(x - x_4) \end{aligned} \tag{3.4.23}$$

$H_5(x)$ can be used to obtain the theoretical length of the unilateral approximate in-plane trocar in a specific ankle joint position. D_3 is influenced by the hip and knee joint angles, illustrated in the $D - H$ coordinate system. The distance from D_3 to the knee joint and from the knee joint axis to the hip joint axis remain constant. By examining changes in hip and knee joint angles at different C_3 and D_3, we can calculate $H_5(x)$ with the maximum arc length as the sheath design length. Additionally, the function $H_5(x)$ provides the full curvature of the unilateral in-plane cannula in a particular ankle joint position.

(4) Path optimization design of multi-joint mechanism

$$\begin{aligned} &H_{n+2}(x) = f[x_1] + f[x_1, x_1](x - x_1) + f[x_1, x_1, x_2](x - x_1)^2 \\ &+ \ldots + f\left[x_1, x_1, x_2, \ldots, x_{n+1}, x_{n+1}\right](x - x_1)^2(x - x_2)(x - x_3)\ldots(x - x_{n+1}) \end{aligned} \tag{3.4.24}$$

When applying the theory to a multi-joint mechanism with in-plane motion using tendon-sheath transmission, the optimization of the tendon-sheath path follows a similar process. For a system with n joints, the path is optimized by approximating it through points A_1, A_2, $A_3 \ldots . A_{n+1}$ and calculating derivatives at A_1, A_{n+1} of the $n + 2$ Newton-type Hermite interpolation polynomial $H_{n+2}(x)$, as given in Eq. (3.4.24). Similarly, the function $H_{n+1}(x)$ derived from Eqs. (3.4.24), combines (3.4.8) and (3.4.14), determines the theoretical length and full curvature of the required tendon-sheath for a specific motion of the end joint in the approximate plane.

References

1. Sun Z, Wang Z, Phee SJ (2013) Elongation modeling and compensation for the flexible tendon-sheath system. IEEE/ASME Trans Mechatron 19(4):1243–1250
2. Hofmann UAT, Bützer T, Lambercy O et al (2018) Design and evaluation of a bowden-cable-based remote actuation system for wearable robotics. IEEE Robot Autom Lett 3(3):2101–2108
3. Chen L (2013) Modeling and control of tendon-sheath transmission system and its application in long and slender robot for searching and inspection. Southeast University
4. Chen L, Wang X, Xu W (2014) Inverse transmission model and compensation control of a single-tendon-sheath actuator. IEEE Trans Ind Electron 61(3):1424–1433
5. Do T, Tjahjowidodo T, Lau M et al (2015) A new approach of friction model for tendon-sheath actuated surgical systems: nonlinear modelling and parameter identification. Mech Mach Theory 85:14–24
6. Kim D, Kim H, Jin S (2022) Recurrent neural network with preisach model for configuration-specific hysteresis modeling of tendon-sheath mechanism. IEEE Robot Autom Lett 7(2):2763–2770
7. Khoshnam M, Patel R (2017) Robotics-assisted control of steerable ablation catheters based on the analysis of tendon-sheath transmission mechanisms. IEEE/ASME Trans Mechatron 22(3):1473–1484
8. Li Y, Lu W, Liu Y et al (2023) Optimization design method of tendon-sheath transmission path under curvature constraint. IEEE Trans Robot
9. Oprea J (2007) Differential geometry and its applications. MAA
10. Li J (2016) Modeling and compliance control of tendon-sheath transmission system. Southeast University
11. Gan Z (2018) The design of a lower limb rehabilitation exoskeleton derived by tendon-sheath artificial muscle. Southeast University

References

Chapter 4
Transmission Model of Tendon-Sheath System

Abstract Tendon-sheath actuation mechanism has been adopted in many robotic applications as it can provide a simple and dexterous remote transmission mode. However, inherent nonlinearities such as backlash and hysteresis arise due to sheath-tendon friction. These nonlinear characteristics, influenced by tendon-sheath configurations, significantly degrade transmission efficiency and control performance, posing challenges in applications demanding high control accuracy. Therefore, establishing accurate transmission characteristic models for tendon-sheath systems is crucial for their robotic applications. This chapter introduce the force and displacement transmission modeling methods of the tendon-sheath systems, including static models and dynamic models. Both the single tendon-sheath transmission mechanism and the double tendon-sheath transmission mechanism are discussed, and the model performance are analyzed by simulation and experiments. The transmission models introduced in this chapter can be applied to the practical applications of the tendon-sheath system to describe the system output, evaluate the transmission capability, and develop precise motion control algorithms, etc.

4.1 Transmission Model of Single Tendon-Sheath System

The tendon-sheath transmission characteristics with a constant curvature have been studied a lot, therefore we shift our attention to the investigation of tendon-sheath systems with arbitrary curves. An arbitrarily bent single-tendon-sheath system and its force balance are illustrated in Fig. 4.1. The distal end of the tendon is connected to a spring with stiffness of k. The sheath has a length L, and the position of a sheath segment is defined by arc length s. We opted for an elastic load because, in practical applications such as the finger joint in [1], the continuum segment in [2], or the brake system of bicycles, elastic components often involved in these single-tendon-sheath systems that serves for one-way actuation. When the tendon pulls the joint or continuum segment, the elastic component deforms and store elastic potential energy. When the tendon is released, the deformation of the elastic component will recover and pull back the joint or continuum segment. The elastic load proves to be a prevalent choice in single-tendon-sheath actuation applications.

X. Wang and Q. Zhang, *Tendon-Sheath Transmitted Light and Slender Robots*,
https://doi.org/10.1007/978-981-96-5581-6_4

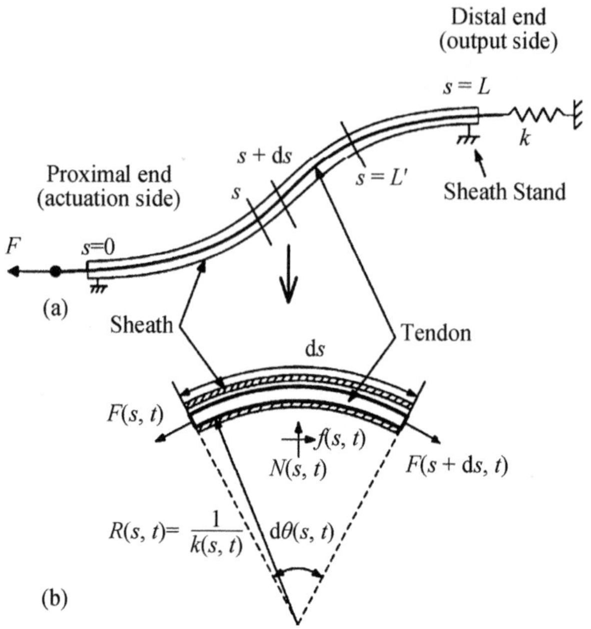

Fig. 4.1 An element of the tendon-sheath transmission mechanism (Reprinted from [3] with permission from IEEE)

To conduct the transmission modeling of the tendon-sheath mechanism, the following assumptions are made [3, 4]:

(1) The mass of the tendon is negligible (the density of the tendon with a diameter of 0.8 mm is only $2.6\ x\ 10^{-3}$ g/mm);
(2) The tendon moves axially along the inner wall of the sheath without lateral movement;
(3) The interaction force between the tendon and the sheath is Coulomb friction;
(4) The deformation of the tendon is within its elastic range and conforms to Hooke's law.

In the differential tendon element, $F(s,t)$ represents input force of tendon element, $F(s+ds,t)$ represents output force of tendon element $f(s,t)$ is friction between tendon and sheath, $R(s,t)$ is radius of the differential tendon element, $d\theta(s,t)$, θ represents curvature angles and s is the arc length. λ and μ represent direction of tendon transmission and friction coefficient respectively, $\lambda = sgn(x_{out})$, x_{out} is the output displacement of tendon-sheath mechanism. The equilibrium equation of the differential tendon element can be established as follows

$$\begin{cases} f(s,t)\lambda + F(s+ds,t)\cos\frac{d\theta}{2} - F(s,t)\cos\frac{d\theta}{2} = 0 \\ F(s,t)\sin\frac{d\theta}{2} + F(s+ds,t)\sin\frac{d\theta}{2} - N(s,t) = 0 \end{cases} \tag{4.1.1}$$

When the tendon-sheath system is divided to be short enough, it exists $d\theta \cong 0$. Ignoring high order infinitesimals, we have

$$\begin{cases} \sin\frac{d\theta}{2} \cong \frac{d\theta}{2} \\ \cos\frac{d\theta}{2} \cong 1 \end{cases} \tag{4.1.2}$$

The equilibrium equation of the differential tendon element can be simplified as:

$$\begin{cases} f(s,t)\lambda + F(s+\mathrm{d}s,t) - F(s,t) = 0 \\ F(s,t)\mathrm{d}\theta - N(s,t) = 0 \end{cases} \tag{4.1.3}$$

With the Coulomb friction model adopted, the basic relations of these variables are shown as follows:

$$\begin{cases} \mathrm{d}F(s,t) = \lambda f(s,t) \\ f(s,t) = \mu N(s,t) \\ N(s,t) = F(s,t)d\theta \end{cases} \tag{4.1.4}$$

In which, $dF(s,t) = F(s+ds,t) - F(s,t)$. From Eq. (4.1.4), the tension distribution can be derived

$$\frac{1}{F(s,t)}\mathrm{d}F(s,t) = -\mu\lambda\mathrm{d}\theta = -\mu\kappa(l)\lambda\mathrm{d}l \tag{4.1.5}$$

In which (l) is the curvature at position s. By integrating over the segment $[s, s + \mathrm{d}s]$, we can get

$$F(s+\mathrm{d}s,t) = F(s,t)e^{\mu\lambda k(s,t)\mathrm{d}s} \tag{4.1.6}$$

Then, we can find the tension in the tendon at arbitrary position s if the input tension $F(0,t)$ is known

$$F(s,t) = F(0,t)e^{\mu\lambda\int_0^s k(\sigma,t)\mathrm{d}\sigma} \tag{4.1.7}$$

Consequently, the force at the distal end is related to the one at the proximal end by

$$F(L,t) = F(0,t)e^{\mu\lambda\int_0^L k(\sigma,t)\mathrm{d}\sigma} \tag{4.1.8}$$

When the whole tendon is stationary, the force at position L is constant, i.e., $F(L,t) = F(L,t^-)$. Thus, the static force transmission model of the single tendon-sheath can be expressed as

$$F(L,t)=\begin{cases}F(0,t)e^{\mu\lambda\int_0^L k(\sigma,t)d\sigma},\lambda\neq 0\\ F\left(L,t^-\right),\lambda=0\end{cases}\tag{4.1.9}$$

Note that, if the sheath does not change its shape, the integration of $k(\sigma,t)$ along its length is constant and the forces at the two ends of the tendon are linearly related if $\lambda\neq 0$. We can rewrite Eq. (4.1.9) as

$$F(L,t)=\begin{cases}F(0,t)Q,\lambda\neq 0\\ F\left(L,t^-\right),\lambda=0\end{cases}\tag{4.1.10}$$

where

$$Q=\begin{cases}Q_p=e^{-\mu\int_0^L k(\sigma,t)d\sigma},\ \lambda<0\\ Q_r=e^{\mu\int_0^L k(\sigma,t)d\sigma},\ \lambda>0\end{cases}\tag{4.1.11}$$

in which $0<Q_p<1<Q_r$ and $Q_p\cdot Q_r=1$ with subscripts p and r for the tendon being pulled and released, respectively.

The tendon experiences elongation when tightened, potentially causing a positioning error in motion or force transmission. We assume that this elongation remains within its elastic limit and can be calculated using Hooke's law.

$$\frac{\Delta L}{L}=\frac{F}{EA}\tag{4.1.12}$$

where ΔL, L, F, E, and A are the tendon elongation, original length, tension, elastic modulus, and cross-sectional area, respectively. Combining Eqs. (4.1.7) and (4.1.12), the total elongation of the tendon during the entire movement can be calculated.

$$\Delta L=\int_0^L\frac{F(s,t)}{EA}\mathrm{d}s=\frac{F(0,t)}{EA}\int_0^L e^{\left[\mu\lambda\int_0^s k(\sigma,t)d\sigma\right]}ds\tag{4.1.13}$$

ΔL is always positive since a tendon can only withstand tension, not compression. For compactness, we define

$$\varphi=\begin{cases}\varphi_p=\int_0^L e^{\left[-\mu\int_0^s k(\sigma,t)d\sigma\right]}ds,\lambda<0\\ \varphi_r=\int_0^L e^{\left[\mu\int_0^s k(\sigma,t)d\sigma\right]}ds,\lambda>0\end{cases}\tag{4.1.14}$$

Let $x(0,t)$ and $x(L,t)$ be the displacements for the two ends of the tendon, then we have

$$x(L,t) = x(0,t) - \Delta L. \tag{4.1.15}$$

At the distal end, we have

$$kx(L,t) = F(L,t) = QF(0,t). \tag{4.1.16}$$

Combining Eqs. (4.1.14)–(4.1.16), we get

$$x(L,t) = \frac{x(0,t)}{1 + \frac{k\varphi}{EAQ}} \tag{4.1.17}$$

The transmission relationship between the positions at the two ends can be expressed as

$$x(L,t) = \begin{cases} Mx(0,t), & \lambda < 0 \\ x\left(L,t^{-}\right), & \lambda \geq 0 \end{cases} \tag{4.1.18}$$

where

$$M = \begin{cases} M_p = \frac{1}{1+\frac{k\varphi_p}{EAQ_p}}, & \lambda < 0 \\ M_r = \frac{1}{1+\frac{k\varphi_r}{EAQ_r}}, & \lambda > 0 \end{cases} \tag{4.1.19}$$

From Eqs. (4.1.18) and (4.1.19), the position transmission is also linear when the whole tendon moves and does not change its shape.

If the sheath has a constant curvature, with the relationship $\theta = L/R = Lk$, we can get

$$Q = \begin{cases} Q_p = e^{-\mu\theta}, & \lambda < 0 \\ Q_r = e^{\mu\theta}, & \lambda > 0 \end{cases} \tag{4.1.20}$$

$$M = \begin{cases} M_p = \frac{1}{1+\frac{kL\left(1-e^{-\mu\theta}\right)}{\theta EAe^{-\mu\theta}}}, & \lambda < 0 \\ M_r = \frac{1}{1+\frac{kL\left(e^{\mu\theta}-1\right)}{\theta EAe^{\mu\theta}}}, & \lambda > 0 \end{cases} \tag{4.1.21}$$

From Eq. (4.1.20), it is clear that the force transmission characteristics are mainly determined by the bending angle of the sheath and the friction coefficient. Note that is the total bending angle of the sheath. From Eq. (4.1.21), we can know that, in addition to the total bending angles, the displacement transmission characteristics are also affected by distal load, original tendon length, tendon elastic modulus, and cross-sectional area. The gap between the tendon and sheath won't affect the transmission characteristics since the tendon is always in tension and slides along the convex surface inside the sheath even when the changing trend of the pulling force is reversed.

4.1.1 Model Simulation

According to the force and displacement transmission model, we carry out the simulation of the transmission model in this section. Assuming that the sheath has constant curvature and its total bending angle is π. The coefficient of friction between the sheath and the tendon is 0.15, the diameter of the tendon is 0.8mm, and the Young's modulus is 1.4×10^5 N/mm^2. Apply the force shown in Fig. 4.2 to the system input, we can get the simulated output of force and displacement as shown in Figs. 4.3 and 4.4, respectively.

From Figs. 4.3 and 4.4, we can see that the transmission of tendon-sheath mechanism is mainly composed of four phases: (A) tightening phase; (B) tightening to relax phase; (C) relax phase; (D) relax to tightening phase. The transmission characteristics of both force and displacement are non-linear, and the loss of force is very obvious in phase A; in the relax phase C, the output force drops rapidly as the input force decreases. During phase B, the output force remains unchanged because only partial tendon moves at this time, and the others remain stationary. In this case, the input changes cannot transmit to the output immediately. The displacement transmission has similar behaviors to the force transmission, except that the slope of the displacement transmission is less than 1 all the time.

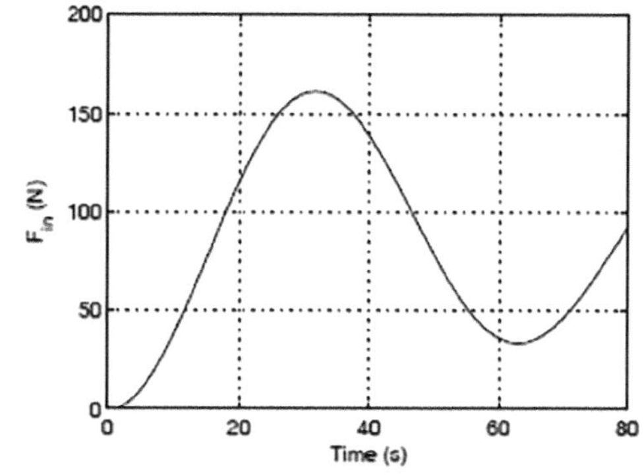

Fig. 4.2 The input force for simulation [12]

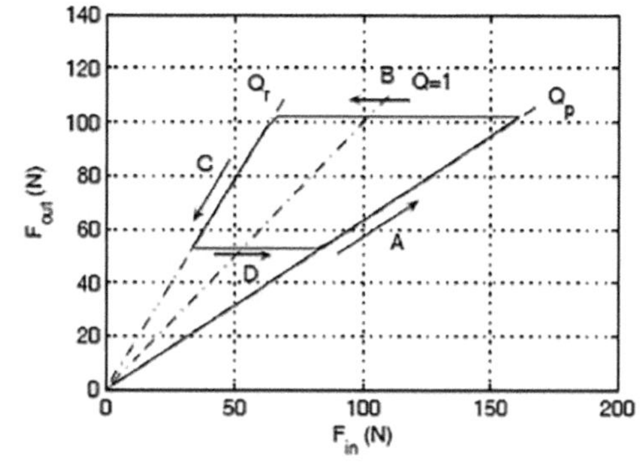

Fig. 4.3 Simulation results of force transmission [12]

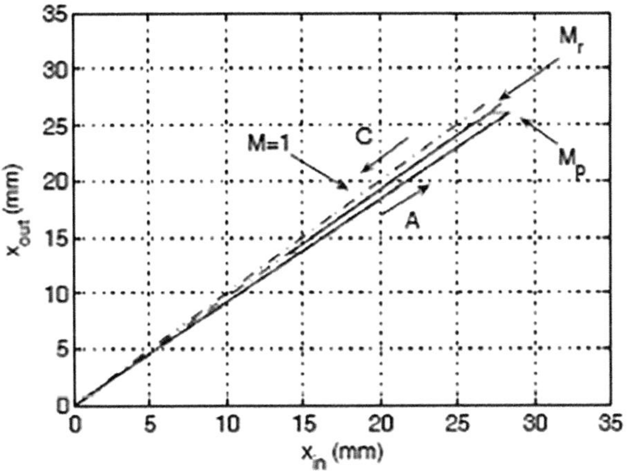

Fig. 4.4 Simulation results of displacement transmission [12]

4.1.2 Experiment Verification

(1) Experiment platform setup

The tendon-sheath transmission model is validated on the experimental platform shown in Fig. 4.5. The two ends of the outer sheath are fixed to a plate and the sheath is winded to a circular plate. The distal end of the tendon is attached to a spring load and the proximal end is connected to a ball screw stage, the stage is driven by a servo motor and runs in speed control mode. The forces at both ends of the tendon are measured by load cells (JLBS-10Kg), and the positions at the stage and spring are measured by position sensors (KTS-75). The signals of the sensors are acquired and processed via a data acquisition card (PCI-1710HG). The tendon used in the experiment is stainless steel wire and has a diameter of 0.8mm. The sheath is a packed spring coil made from steel wire with a 0.5 mm diameter and has an inner diameter of 1.2 mm. The sheath can withstand a significant amount of axial force. The lengths of the sheath and tendon are 2 m and 2.2 m, respectively.

(2) Model parameters identification

To obtain the constants Q_p, Q_r, M_p, and M_r of the tendon-sheath transmission model, we must integrate the curvature (σ, t) over the sheath segment. However, the exact (σ, t) is hard to know when the tendon-sheath mechanism is in action. In what follows, an experimental method is presented to identify the values of Q_p, Q_r, M_p, and M_r using sensors at the proximal end of the mechanism.

As shown in Fig. 4.6a, a stopping component is used to connect the tendon at the distal end, where the load and position sensors are also placed. In the beginning, the

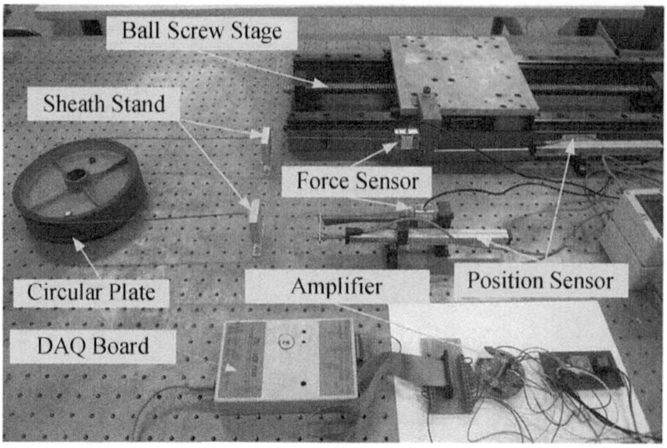

Fig. 4.5 Schematic of the experimental setup (Reprinted from [3] with permission from IEEE)

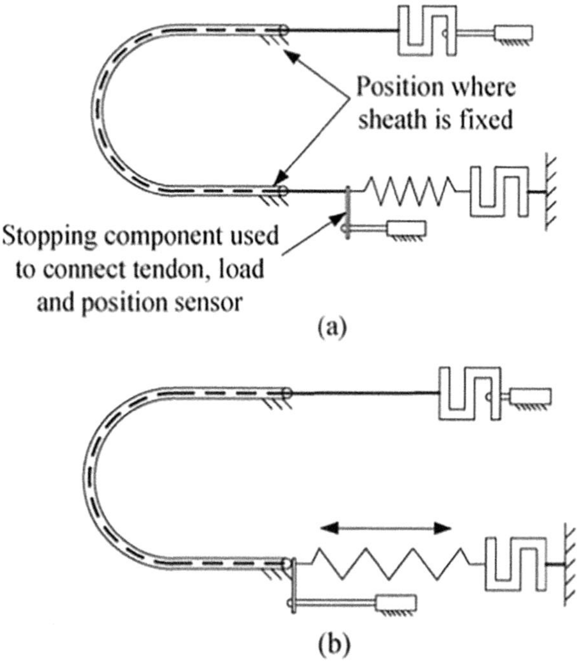

Fig. 4.6 Parameter identification procedure (Reprinted from [3] with permission from IEEE)

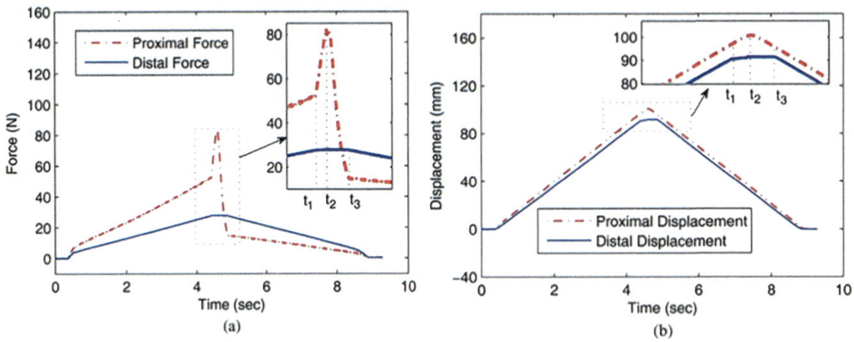

Fig. 4.7 Measured positions and forces at two ends of the tendon (Reprinted from [3] with permission from IEEE)

ball screw stage is actuated to pull the tendon until the stopping component meets the sheath end fixed at a site, and afterward, the stage moves inversely back to its original place in the end.

During this process, the measured forces and displacements at both the proximal side and distal side are shown in Fig. 4.7, where t_1 is the moment when the stopping component comes in touch with the sheath end, t_2 is the moment when the ball screw stage actuation is reversed in direction, and t_3 is the time when the stopping component departs from the sheath end. The force at the actuation side (proximal) increases sharply after the stopping component has met the sheath end, between t_1 and t_2. The output force and position are constant, between t_1 and t_3, before the stopping component loses contact with the sheath end, even though the actuation is reversed at t_2. This is because the deformation that occurred between t_1 and t_2 takes time to be recovered after the actuation direction is reversed at the proximal end. From t_3, the deformation is fully recovered, and the proximal and distal sides change simultaneously again.

According to Eq. (4.1.10), we have the relationship

$$F(L, t_1) = F(0, t_1)\, Q_p \tag{4.1.22}$$

$$F(L, t_3) = F(0, t_3)\, Q_r \tag{4.1.23}$$

Since $F(L, t)$ is constant between t_1 and t_3, we have

$$F(0, t_1)\, Q_p = F(0, t_3)\, Q_r \tag{4.1.24}$$

Considering the relationship $Q_p Q_r = 1$, we can get

$$Q_p = \sqrt{\frac{F(0, t_3)}{F(0, t_1)}} \tag{4.1.25}$$

As long as the stopping component is in touch with the sheath end, the displacement of the ball screw stage is due entirely to the deformation of the tendon within the sheath. From Eq. (4.1.13), between t_1 and t_2, we have

$$\mathrm{x}\left(0, t_2\right) - \mathrm{x}\left(0, t_1\right) = \frac{F\left(0, t_2\right)}{EA}\varphi_p - \frac{F\left(0, t_1\right)}{EA}\varphi_p \tag{4.1.26}$$

Such that we can calculate the φ_p by

$$\varphi_p = \frac{EA\left[x\left(0, t_2\right) - x\left(0, t_1\right)\right]}{F\left(0, t_2\right) - F\left(0, t_1\right)} \tag{4.1.27}$$

After the ball screw stage changes its actuation direction, between t_2 and t_3, we have

$$\mathrm{x}\left(0, t_3\right) - \mathrm{x}\left(0, t_2\right) = \frac{F\left(0, t_3\right)}{EA}\varphi_r - \frac{F\left(0, t_2\right)}{EA}\varphi_p \tag{4.1.28}$$

Such that the φ_r can be calculated as

$$\varphi_r = \frac{EA\left[x\left(0, t_3\right) - x\left(0, t_2\right)\right]}{F\left(0, t_3\right)} + \frac{F\left(0, t_2\right)}{F\left(0, t_3\right)}\varphi_p \tag{4.1.29}$$

To validate the identification method for arbitrary sheath shape, this method was tested by the sheath with a known shape. The sheath was curved to bending angles of $\pi/2$, π, and $3\pi/2$, respectively, around a circular plate with a radius of 72 mm. For each bending angle, the identification process was repeated five times. With the measurements at the proximal side, the mean value and standard deviation (M. V. and S. D., for short) of identified Q_p, Q_r, M_p, and M_r are shown in Table 4.1. Moreover, these results are compared with the theoretical values, and the maximal relative error (R.E., for short) is only 5.92%.

Table 4.1 Identified parameters for known sheath shape (Reprinted from [3] with permission from IEEE)

Bending angle	Parameters	Q_p	Q_r	M_p	M_r
$\frac{\pi}{2}$	M. V.	0.746	1.3406	0.9748	0.9787
	S. D.	0.0009	0.0019	0.0065	0.007
	R. E.	5.58%	5.92%	2.35%	2%
π	M. V.	0.6306	1.5854	0.9782	0.9805
	S. D.	0.0055	0.0162	0.0037	0.0054
	R. E.	1.03%	1.04%	1.8%	1.7%
$\frac{3\pi}{2}$	M. V.	0.4802	2.082	0.9503	0.9511
	S. D.	0.0069	0.0296	0.0047	0.0045
	R. E.	2.64%	2.68%	4.35%	4.58%

Then, the circular plate was removed, and the sheath was curved to an arbitrary shape with unknown curvature distributions. The mean values of Q_p, Q_r, M_p, and M_r and their standard deviations are shown in Table 4.2. The identified parameters and experimental data are plotted in Figs. 4.8 and 4.9, where we can see that the identified slopes are in good agreement with the experimental results. It is noted that, as shown in Fig. 4.9, the experimental profile has a small arc rather than a straight line in the beginning, since the tendon is not fully tightened and, consequently, does not apply any tension.

Table 4.2 Identified parameters for unknown sheath shape (Reprinted from [3] with permission from IEEE)

Parameters	Q_p	Q_r	M_p	M_r
Mean value	0.4915	2.035	0.949	0.9904
Standard deviations	0.0058	0.0161	0.0198	0.0007

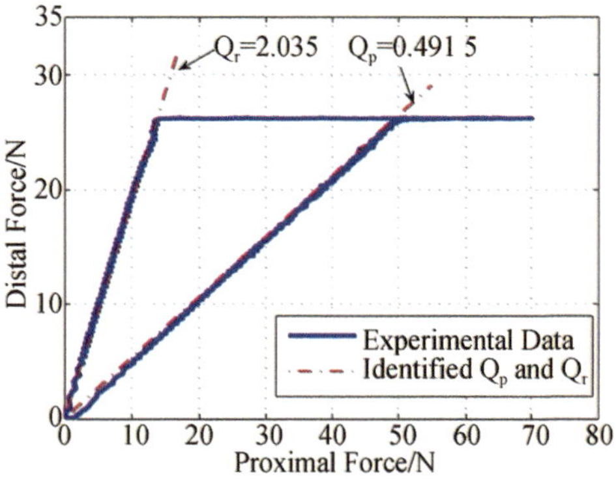

Fig. 4.8 Identified force transmission slopes Q_p and Q_r and the experimental force transmission profile (Reprinted from [3] with permission from IEEE)

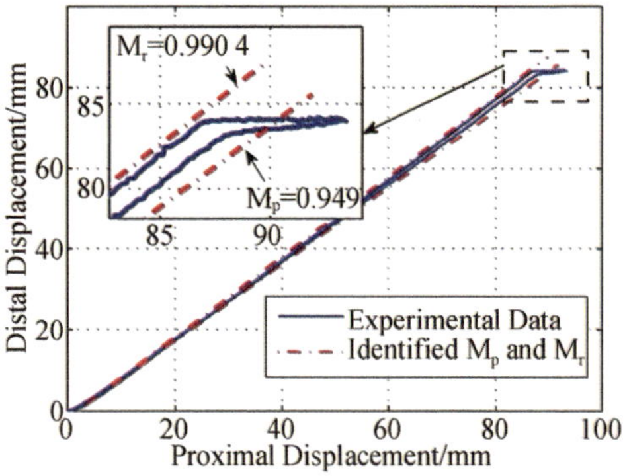

Fig. 4.9 Identified position transmission slopes M_p and M_r and the experimental position transmission profile (Reprinted from [3] with permission from IEEE)

(3) Model verification

The transmission model is tested by several experiments with different setup, including the bending angle of the sheath, the bending radii and the movement speed. The experimental results are presented in Figs. 4.10–4.12.

Case 1: Variation of bending angles According to the transmission model, both the force and displacement transmission characteristics are related to the total bending angle of the sheath. To verify this, the particular bending angles of the sheath are set to $\pi/2$, π, and $3\pi/2$ by the circular plate with a constant bending radius of 72 mm. The force and displacement transmission results were shown in Fig. 4.10a and b. We can see that the increase in bending angle will cause larger force and displacement transmission nonlinearities. Moreover, the experimental data match well with the theoretical values. For clarity, only two straight lines with slopes of M = 0.9506 and M = 0.995, which are the theoretical values of M_p and M_r with a bending angle of $3\pi/2$, are marked in Fig. 4.10b.

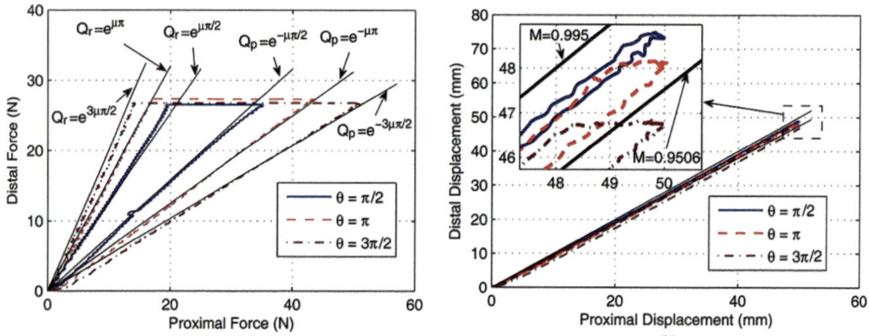

(a) (b)

Fig. 4.10 Transmission characteristics with different bending angles (Reprinted from [3] with permission from IEEE)

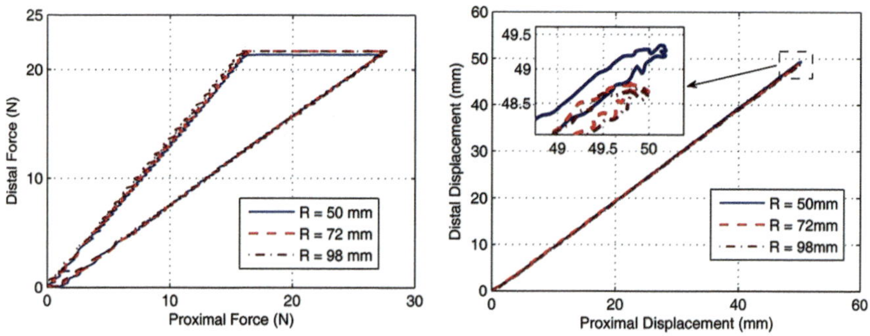

(a) (b)

Fig. 4.11 Transmission characteristics with different bending radii (Reprinted from [3] with permission from IEEE)

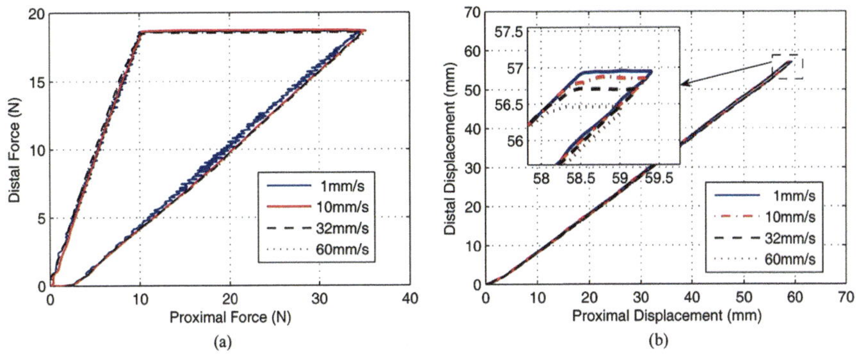

Fig. 4.12 Transmission characteristics with different motion velocities (Reprinted from [3] with permission from IEEE)

Case 2: Variation of bending radii In this experiment, different circular plates with radii of 50mm, 72mm, and 98 mm were used while the sheath was always wrapped to a constant bending angle of π. From Fig. 4.11a, we can know that the variation of bending radii almost has no effect on force transmission. As $L = R\theta$, the change of radius only results in a small change in tendon elongation, as Fig. 4.11b shows, it only brings a minor effect in displacement transmission.

Case 3: Variation of motion velocity In this experiment, the sheath has a bending angle of π and a radius of 98 mm, and the stage moves at different velocities. As the velocity increases, the force transmission characteristics have no obvious change, as shown in Fig. 4.12a. The transmission profile has some chattering when the velocity is very low, and this may be caused by the stick-slip of the stage. Figure 4.12b shows that a very high velocity will cause large tendon elongation. However, the tendon-sheath actuation system works at a low velocity in most practical scenarios, e.g., robotic hand and surgical robot; thus, this effect of the variation of velocity can be ignored.

4.2 Transmission Model of Double Tendon-Sheath System

Figure 4.13 shows a schematic diagram of a robot joint actuated by a remote motor through a double-tendon-sheath transmission system [5]. Tendons a and b are attached to the pulleys on the input side and output side in pull-pull configuration. The pulling force is transmitted by mechanical displacement between inner tendon and outer sheath. r_1 and r_2 are the radiuses of each pulley. θ_{in}, θ_{out}, ω_{in}, and ω_{out} are the rotary displacements and velocities of driving motor and robot joint. T_0 is the pretension of the system, which can be adjusted by a pretension device. T_1, T_2, T_3, and T_4 denote the tensions at the four ends of the tendons. τ_{in} is the input torque applied by the motor and τ_{out} is the output torque acting on the robot joint. In the double-tendon-sheath actuation analysis, it is assumed that the pretension in the

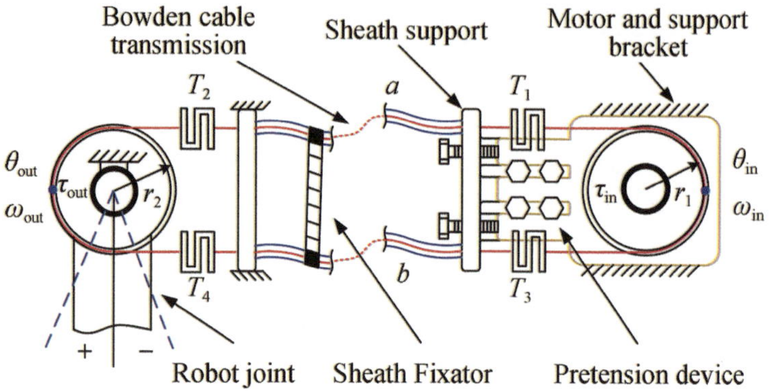

Fig. 4.13 Schematic diagram of double-tendon-sheath transmission system (Reprinted from [5] with permission from IEEE)

tendon is large enough to prevent any system slacking due to tendon elasticity. Here, clockwise is defined as the positive direction for the rotary displacements, velocities, and torques.

For modeling simplicity, the curvature distribution along the longitudinal direction is assumed to remain the same for each tendon during the operation of robot joint. This assumption is rational since the sheaths can be connected to each other with several fixing parts, as shown in Fig. 4.13. With the integral of the curvature to the whole arc-length defined as the total sheath bending angle θ, we have

$$\theta = \int_0^L k(\sigma, t)\mathrm{d}\sigma \tag{4.2.1}$$

Meanwhile, the total bending angle can be expressed as the product of the average curvature κ and the tendon length L, i.e., $\theta = \kappa L$. In fact, it is a valuable related to the sheath configuration.

According to expressions of tension and elongation in Eqs. (4.1.10) and (4.1.13), $T_{out}(t)$ and ΔL keep their previous values when the tendon stays relatively static to the sheath. As a result, the whole transmission system operates in a balanced state as well. It is thus clear that what we need is to derive the transmission model while the system goes in the state of motion. So we can rewrite Eqs. (4.1.10) and (4.1.13) into

$$T_{out}(t) = T_{in}(t)e^{\lambda\mu\theta} \tag{4.2.2}$$

$$\Delta L(t) = \frac{T_{\mathrm{in}}(t)}{AE}\int_0^L e^{-\lambda\mu\kappa s}ds = -\frac{T_{\mathrm{in}}(t)(e^{-\lambda\mu\theta} - 1)}{\lambda AE\mu\kappa}. \tag{4.2.3}$$

Apparently, the tensions at the two ends of the tendon, as calculated by Eq. (4.2.2), are linearly related if the sheath keeps its shape as well as its fixed curvature distribution. As the tendon can only withstand tension but not compression, the elongation, which is a time varying function of input tension, turns out to be always positive, as calculated by Eq. (4.2.3).

When the actuation system rotates clockwise, the force transmission relationship at each tendon end can be expressed as follows:

$$T_2 = e^{-\mu\theta} T_1 = Q_p T_1 \tag{4.2.4}$$

$$T_4 = e^{\mu\theta} T_3 = Q_r T_3. \tag{4.2.5}$$

where $Q_p = e^{-\mu\theta}$ and $Q_r = e^{\mu\theta}$. It is worth noting that Q_p and Q_r are not constant, as the sheath may change its geometric configuration during operation.

In the torque balancing equation, the input and output torque can be represented as the product of the tension difference and the radius of the corresponding pulley:

$$\tau_{in} = r_1(T_1 - T_3) \tag{4.2.6}$$

$$\tau_{in} = r_2(T_2 - T_4) \tag{4.2.7}$$

Assume that the elongation keeps within elastic limitation. The initial angle values of θ_{in} and θ_{out} are equivalent in this system. Since the pretension in the tendon is large enough to eliminate slacking, the elongations of tendon a and b can be calculated by

$$\Delta L_a = r_1\theta_{\text{in}} - r_2\theta_{\text{out}} \tag{4.2.8}$$

$$\Delta L_a + \Delta L_b = 0. \tag{4.2.9}$$

While the entire tendon segment keeps moving, by substituting Eqs. (4.2.3), (4.2.4) and (4.2.5) into (4.2.8) and (4.2.9), the elongation can be refined as

$$\Delta L_a = \frac{1}{AE}\int_0^L \left(T_1 e^{-\mu\kappa s} - T_0\right) ds = \frac{T_1\psi_p - T_0 L}{AE} \tag{4.2.10}$$

$$\Delta L_b = \frac{1}{AE}\int_0^L \left(T_3 e^{\mu\kappa s} - T_0\right) ds = \frac{T_3\psi_r - T_0 L}{AE} \tag{4.2.11}$$

where $\psi_p = \frac{L}{\mu\theta}(1 - Q_p)$, $\psi_r = \frac{L}{\mu\theta}(1 - Q_r)$.

Base on Eqs. (4.2.10) and (4.2.11), the relationship between the tensions on the drive side can be expressed as

$$T_1\psi_p - T_3\psi_r = 2T_0L \tag{4.2.12}$$

Combining Eqs. (4.2.4), (4.2.5), (4.2.6) and (4.2.12), the tensions acting on the output side are related to the input torque by the following equations:

$$T_2 = \frac{Q_p\left(\psi_r\tau_{in} - 2T_0Lr_1\right)}{r_1\left(\psi_r - \psi_p\right)} \tag{4.2.13}$$

$$T_4 = \frac{Q_r\left(\psi_p\tau_{in} - 2T_0Lr_1\right)}{r_1\left(\psi_r - \psi_p\right)} \tag{4.2.14}$$

Substituting Eqs. (4.2.13) and (4.2.14) into (4.2.7) yields the torque transmission relationship between the two system ends as follows:

$$\begin{aligned}\tau_{\text{out}} &= \frac{r_2\left[\left(\psi_r\tau_{\text{in}} - 2T_0Lr_1\right)Q_p - \left(\psi_p\tau_{\text{in}} - 2T_0Lr_1\right)Q_r\right]}{r_1\left(\psi_r - \psi_p\right)} \\ &= \frac{\tau_{\text{in}}r_2}{r_1} - 2T_0r_2\mu\theta.\end{aligned} \tag{4.2.15}$$

Inserting Eqs. (4.2.10) and (4.2.13) into (4.2.8) gives the displacement transmission relational expression Eq. (4.2.16).

$$\theta_{\text{out}}(t) = \frac{r_1}{r_2}\theta_{\text{in}}(t) - \frac{\tau_{\text{in}}(t)\psi_r\psi_p - T_0Lr_1\left(\psi_r + \psi_p\right)}{AEr_1r_2\left(\psi_r - \psi_p\right)}. \tag{4.2.16}$$

In the same way, when the actuation system rotates counterclockwise, the relationship between torque and displacement transmission can be expressed as

$$\tau_{\text{out}} = \frac{\tau_{\text{in}}r_2}{r_1} + 2T_0r_2\mu\theta \tag{4.2.17}$$

$$\theta_{\text{out}}(t) = \frac{r_1}{r_2}\theta_{\text{in}}(t) - \frac{\tau_{\text{in}}(t)\psi_r\psi_p + T_0Lr_1\left(\psi_r + \psi_p\right)}{AEr_1r_2\left(\psi_r - \psi_p\right)}. \tag{4.2.18}$$

Combining Eqs. (4.2.15) and (4.2.17) and considering the situation when the system is in motionless state, we obtain the final expression for torque transmission, i.e.,

$$\tau_{\text{out}}(t) = \begin{cases} \frac{\tau_{\text{in}}(t)r_2}{r_1} - \operatorname{sign}(\omega_{\text{in}})2T_0r_2\mu\theta, & \omega_{\text{in}} \neq 0 \\ \tau_{\text{out}}\left(t^-\right), & \omega_{\text{in}} = 0. \end{cases} \tag{4.2.19}$$

As indicated by Eq. (4.2.19), the actual torque transmission model is mainly determined by pulley radius, system pretension, total bending angle, friction coefficient, and the direction of rotation. The dynamic resisting torque caused by the friction between tendon and sheath is $-sign(\omega_{in})2T_0r_1\mu\theta$. If $|\tau_{\text{in}}\, r_2/r_1 - \tau_{\text{out}}| \leq 2T_0r_1\mu\theta$, the system will remain stationary, the output torque being constant. If $|\tau_{\text{in}}\, r_2/r_1 - \tau_{\text{out}}| > 2T_0r_1\mu\theta$, the system will overcome the frictional torque and get into a motion state.

Equations (4.2.16) and (4.2.18) show the displacement transmission relationship when both the pulleys are moving in the same direction. The rotary displacement transmission profiles are of backlash-like characteristics. When the direction of the driving pulley is altered, the motion changing at the actuation side cannot be transmitted to the output pulley instantly by either of the tendons due to the elastic elongation. Consequently, the output position will remain unchanged until the backlash offset is compensated and all the tendon elements are in moving state. For simplification, we define

$$\varphi(t)\begin{cases}\varphi_s(t) = \frac{\tau_{\text{in}}(t)\psi_r\psi_p - T_0Lr_1(\psi_r+\psi_p)}{AEr_1r_2(\psi_r-\psi_p)}, & \omega_{\text{in}}(t) > 0\\ \varphi_n(t) = \frac{\tau_{\text{in}}(t)\psi_r\psi_p + T_0Lr_1(\psi_r+\psi_p)}{AEr_1r_2(\psi_r-\psi_p)}, & \omega_{\text{in}}(t) < 0.\end{cases} \tag{4.2.20}$$

Let t_0 denote the last time when the system enters the backlash phase, $\Delta\theta_{in}$ is the remaining backlash on the input side. Since the output position is invariable during the backlash compensation process, we have

$$\theta_{\text{out}}(t_0) = \frac{r_1}{r_2}\theta_{\text{in}}(t_0) - \varphi(t_0) \tag{4.2.21}$$

$$\theta_{\text{out}}(t) = \frac{r_1}{r_2}\theta_{\text{in}}(t) - \varphi(t) \tag{4.2.22}$$

$$\Delta\theta_{\text{in}}(t) = \frac{r_2}{r_1}[\varphi(t) - \varphi(t_0)] = \begin{cases}\varphi_s(t) - \varphi_n(t_0), & \omega_{\text{in}}(t) > 0\\ \varphi_n(t) - \varphi_s(t_0), & \omega_{\text{in}}(t) < 0.\end{cases} \tag{4.2.23}$$

Once $\Delta\theta_{in}$ goes negative, the backlash offset is totally compensated, and then we obtain

$$\theta_{\text{out}}(t) = \begin{cases}\frac{r_1}{r_2}\theta_{\text{in}}(t) - \varphi_s(t), & \omega_{\text{in}}(t) > 0 \text{ and } \Delta\theta_{\text{in}}(t) < 0\\ \frac{r_1}{r_2}\theta_{\text{in}}(t) - \varphi_n(t), & \omega_{\text{in}}(t) < 0 \text{ and } \Delta\theta_{\text{in}}(t) < 0\\ \theta_{\text{out}}(t^-), & \text{other else.}\end{cases} \tag{4.2.24}$$

As revealed by Eqs. (4.2.20) and (4.2.24), in addition to pulley radius, system pretension, total bending angle, friction coefficient and rotation direction, the displacement transmission are also affected by tendon length, elastic modulus, cross-sectional area and the input torque. $\varphi(t)$ is the position error triggered by the tendon elasticity and, therefore, needs to be compensated in the motion control. The stiffness of the sheath is neglected here in that all the sheaths are fixed to the support and thus suffer no tendon counterforce.

It is worth noting that the parameters in Eqs. (4.2.19) and (4.2.24) are not concerned with the specific load expression on the output side. Therefore, the proposed transmission model is suitable for any type of environmental load, differing from the previous works on the double-tendon-sheath actuation system.

4.2.1 Experiment Verification

(1) Model parameters identification

In terms of the aforementioned transmission models, most of the parameters (i.e., r_1, r_2, L, E, and A) in Eqs. (4.2.19) and (4.2.24) are constants, and thus obtainable before the actuation system is installed. The variables, such as θ_{in}, ω_{in}, and τ_{in} can be measured using the encoders and force sensors placed on the actuation side. However, it should be noted the product of the total sheath bending angle θ and the friction coefficient μ is the key point for model identification, as the precise value of θ is hard to determine with the actuator in action and, what is more, μ varies with the change of the lubrication conditions between tendon and sheath. In what follows, an experimental method is proposed for the identification of $\mu\theta$.

The schematic of parameters identification is presented in Fig. 4.14. A stopping component is used to limit the maximum output angle θ_{max} at the distal end. θ_{max} becomes known once the overall structure is determined. At the beginning, the output robot joint is driven to rotate clockwise around the fixed shaft until it meets the stopping component at the time t_s, and then the actuation system needs to keep itself in static equilibrium. The forces and position information are required for model identification.

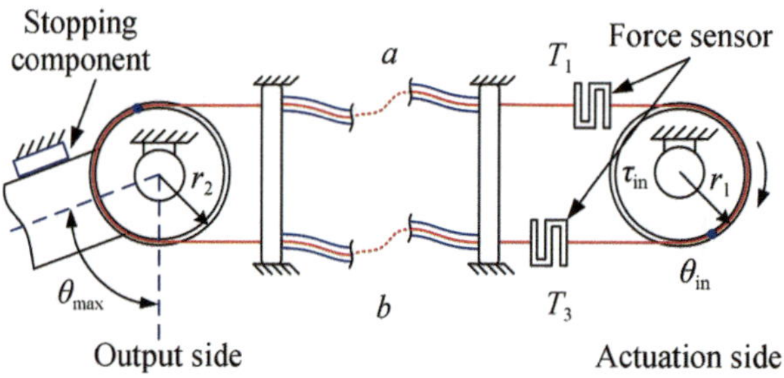

Fig. 4.14 The schematic of parameters identification (Reprinted from [5] with permission from IEEE)

From Eqs. (4.2.8) and (4.2.10), we can get:

$$\begin{aligned}\Delta L_a(t_s) &= r_1\theta_{\text{in}}(t_s) - r_2\theta_{\max} \\ &= \frac{1}{AE}\left[\frac{T_1(t_s)}{\mu\kappa} - \frac{T_1(t_s)}{\mu\kappa}Q_p - T_0L\right]\end{aligned} \tag{4.2.25}$$

Equation (4.2.25) can be converted into Eq. (4.2.26) to express the relationship of $\mu\theta$:

$$\frac{AE\left[r_1\theta_{\text{in}}(t_s) - r_2\theta_{\max}\right] + T_0L}{T_1(t_s)L} = \frac{\left(1 - e^{-\mu\theta}\right)}{\mu\theta}. \tag{4.2.26}$$

For compactness, let us define

$$f(t_s) = \frac{AE\left[r_1\theta_{\text{in}}(t_s) - r_2\theta_{\max}\right] + T_0L}{T_1(t_s)L}. \tag{4.2.27}$$

Thus, Eq. (4.2.26) can be rewritten as

$$e^{-\mu\theta} + f(t_s)\mu\theta - 1 = 0. \tag{4.2.28}$$

As the last step, the Newton-iterative law is adopted to estimate the value of $\mu\theta$ using MATLAB optimization toolbox from MathWorks, Inc. This off-line identification method is useful for the tendon-sheath robot that has to work in confined space with strict spatial constraints. The robot is shape hardly undergoes any change once it reaches its working site. However, this method is inapplicable to cases with frequent sheath configuration changes during the working process.

(2) Torque transmission model verification

The experimental setup for torque transmission validation is displayed in Fig. 4.15. The platform includes a servo AC motor (YASKAWA-SGMAV, rate torque-1.75

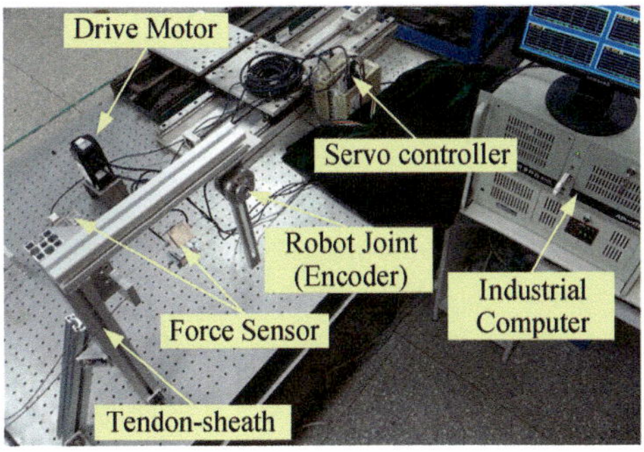

Fig. 4.15 Experimental setup for torque transmission (Reprinted from [5] with permission from IEEE)

Nm) as the actuator, a robot joint made of duralumin material as the environmental load, and a double-tendon-sheath transmission system. The tendons used in the experiments are stainless steel wires with a diameter of 0.8 mm. The outer sheaths are tightly wound springs made from the 0.5 mm steel wire and have an inner diameter of 1.2 mm. The sheaths and tendons are 1 m and 1.2 m in length. The proximal and distal ends of tendons are attached to the pulleys at the actuation side and load side in a pull-pull configuration. The robot joint is installed in precision bearings in order to reduce rolling friction. The total bending angle of sheath is set to rad. The diameters of the pulleys are both 40 mm. The pretension of the system is set at approximately 30 N. The rotary displacement of motor is measured by its own encoder, and the forces at both ends of the tendons are measured by load cells (JLBS-10Kg). The signals of the sensors, amplified by a transistor made amplifier, are acquired and processed via a data acquisition card (PCL-818, Advantech). A real-time control system is established in xPC target environment based on Matlab/RTW. Two industrial personal computers (IPC-610H, Advantech) are used as the host and slave computers.

The servo motor was run in torque control mode with a sinusoidal wave motion trajectory of $\tau_{in} = 0.8sin(\pi/5t)$ Nm. The theoretical values of output torque were obtained using Eq. (4.2.19). And the actual output torque was measured by the force sensors placed at the distal end. Figure 4.16 shows the comparison between the simulation and experiment results of the output torque during a period of 10 s, which turn out to be quite consistent. For better understanding of the model, the torque transmission process can be divided into the following phases:

A: τ_{in} begins to increase in the positive direction, and satisfies the criteria $\tau_{in}r_2/r_1 \le 2T_0 r_2\mu\theta$. The actuator fails to overcome the frictional resistance so that the system remains stationary, $\tau_{out} = 0$.

B: τ_{in} keeps increasing in the positive direction, and satisfies the criteria $\tau_{in}r_2/r_1 > 2T_0 r_2\mu\theta$. The actuator overcomes the frictional resistance, making the system rotating clockwise, $\tau_{out} = \tau_{in}r_2/r_1 - 2T_0 r_2\mu\theta$.

C: τ_{in} begins to decrease in the negative direction, and satisfies the criteria $\tau_{out} - \tau_{in}r_2/r_1 \le 2T_0 r_2\mu\theta$. The actuator fails to overcome the frictional resistance so that the system remains stationary, τ_{out} remains constant.

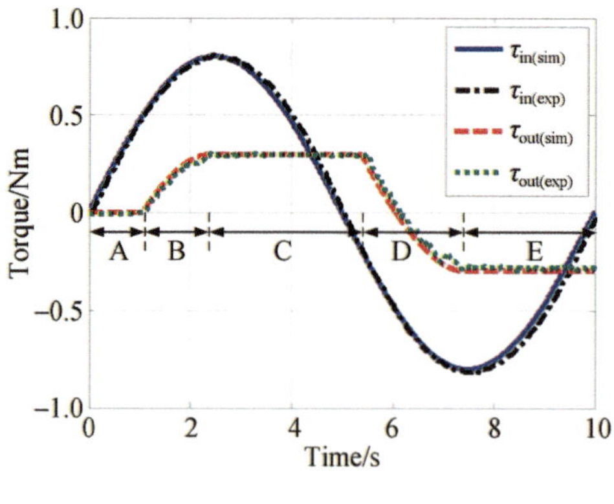

Fig. 4.16 Experimental and simulation results of force transmission (Reprinted from [5] with permission from IEEE)

D: τ_{in} keeps decreasing in the negative direction, and satisfies the criteria $\tau_{out} - \tau_{in}r_2/r_1 > 2T_0r_2\mu\theta$. The load end becomes active and, furthermore, overcomes the frictional resistance, making the system rotating counterclockwise, $\tau_{out} = \tau_{in}r_2/r_1 + 2T_0r_2\mu\theta$.

E: τ_{in} begins to increases in the positive direction, and satisfies the criteria $\tau_{in}r_2/r_1 - \tau_{out} \leq 2T_0r_2\mu\theta$. The actuator fails to overcome the frictional resistance so that the system remains stationary, τ_{out} being constant.

(3) Displacement transmission model verification

The experimental setup for displacement transmission model validation is displayed in Fig. 4.17. For the purpose of providing predefined load torque, the robot joint is replaced by a servo AC motors (YASKAWA-SGMAV, rate torque-1.75 Nm) run in torque control mode as the environmental load. The transmission system and real-time control system are the same as the platform for torque transmission model validation. The drive motor was run in position control mode to track the predefined trajectories. The input and output displacements are measured by the encoders of the servo motors. In the validation experiments, the position command inputted into the driving motor is to move in a sinusoidal wave motion trajectory of $\theta_{in} = \pi/4sin(2\pi/5t)$ rad. Meanwhile, the load torque command for the load motor is $\tau = -0.7sin(\theta_{in})$ Nm. Since the external load condition is known, the theoretical values of output position could be obtained using Eqs.(4.2.20), (4.2.23) and (4.2.24). Figure 4.18 shows the comparison between the simulation and experimental results of the output position during a period of 5 s, which turn out to be quite consistent. The transmission process falls into the following phases:

A: The driving motor begins to rotate clockwise, $\omega_{in}(t) > 0$, while the backlash offset is not completely compensated, $\Delta\theta_{in} > 0$. Thus, the output position remains stationary, $\theta_{out} = 0$.

B: The driving motor keeps rotating clockwise, $\omega_{in}(t) > 0$, and the backlash offset is completely compensated, $\Delta\theta_{in} < 0$. Thus, the output end is driven to rotate clockwise as well, $\theta_{out}(t) = \theta_{in}(t)r_1/r_2 - \varphi_s(t)$.

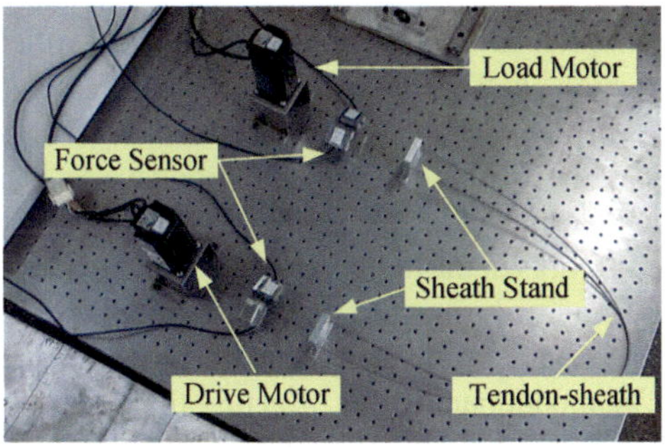

Fig. 4.17 Experimental setup for displacement transmission (Reprinted from [5] with permission from IEEE)

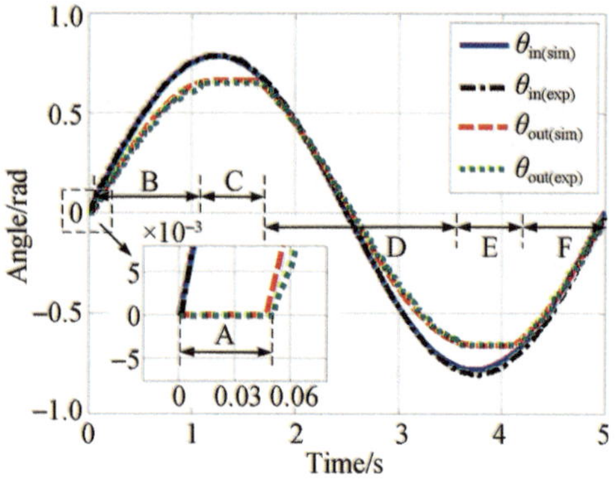

Fig. 4.18 Experimental and simulation result of displacement transmission (Reprinted from [5] with permission from IEEE)

C: The driving motor changes the direction of rotation, $\omega_{in}(t) < 0$, while the backlash offset is not completely compensated, $\Delta\theta_{in} > 0$. Thus, the output position remains stationary, $\theta_{out}(t) = \theta_{out}(t^{-})$.

D: The driving motor keeps rotating counterclockwise, $\omega_{in}(t) < 0$, and the backlash offset is completely compensated, $\Delta\theta_{in} < 0$. Thus, the output end is driven to rotate counterclockwise as well,$\theta_{out}(t) = \theta_{in}(t)r_1/r_2 - \varphi_n(t)$.

E: The driving motor changes the direction of rotation, $\omega_{in}(t) > 0$, while the backlash offset is not completely compensated, $\Delta\theta_{in} > 0$. Thus, the output position remains stationary, $\theta_{out}(t) = \theta_{out}(t^{-})$.

F: Repeat phase B.

4.2.2 Analysis of Transmission Efficiency

The frictional resistance moment suffered by the double tendon-sheath transmission system during operation is the most important factor affecting its transmission efficiency. According to the moment transmission model, it can be seen that the size of the $2F_0r_2\mu\theta$ term must be reduced to improve the system transmission efficiency. Therefore, the transmission efficiency can be improved by reducing the system preload force F_0, reducing the load roller radius r_2, reducing the tendon-sheath friction coefficient μ and reducing the full curvature of the tendon-sheath θ.

In the experimental platform of double tendon-sheath transmission efficiency analysis, the drive motor at the drive end works in the position mode and tracks the motion with a sinusoidal trajectory of $\theta_{in} = \pi sin(0.4\pi t)/4$ rad; the load motor at the load end works in the moment mode, which is used to simulate the load moment generated by gravity during the swing of the robot arm, and the output moment can be described as an expression about the angle of the load roller, i.e.$\tau_{load} = -\sqrt{2}sin\theta_{out}/2\ N \cdot m$. The model parameters affecting the transmission efficiency are adjusted separately, and the experimental results of the torque transmission characteristics are analyzed and discussed.

(1) Change of tendon-sheath preload force

This set of experiments analyzes the variation characteristics of the transmission efficiency by changing the magnitude of the preload force of the system. Using the disk shown in Fig. 4.19 to set the full curvature of the tendon-sheath to π rad, the wire rope and the casing are in a dry friction state, the radius of both the driving roller and the load roller is 20 mm, and the preload force of the system is adjusted to 18 N, 24 N and 30 N respectively by the preload device, and then the torque transmission characteristics analysis experiments are conducted to measure the torque at the driving end and the load end respectively, and the experimental results are shown in Fig. 4.20. It can be seen that under the same conditions of the system's motion law and external load, the larger the preload force of the system, the larger the drive torque output by the drive motor, and therefore the lower the transmission efficiency. The experimental results are reasonable, according to the transmission model, the increase of the system preload will simultaneously lead to the increase of the positive pressure between the tendon-sheath mechanism and the increase of the system frictional resistance. Therefore, excessive system preload should be avoided

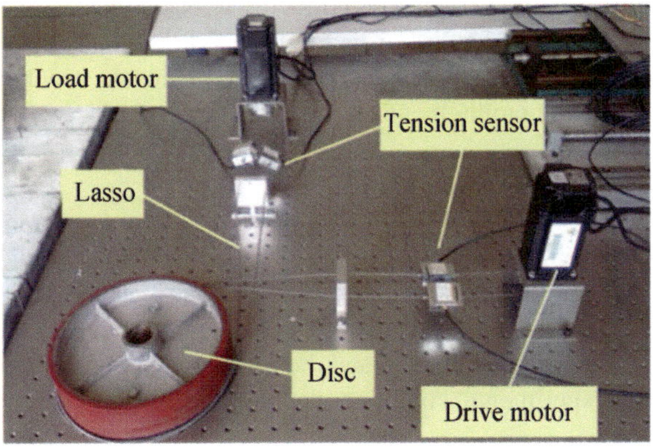

Fig. 4.19 Physical drawing of the experimental platform [4]

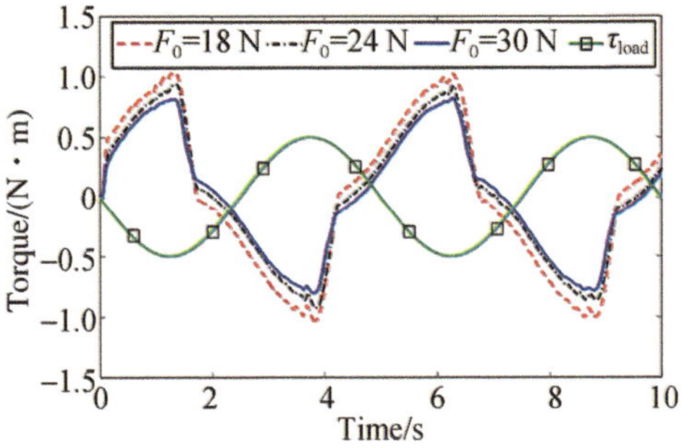

Fig. 4.20 Change in preload force [4]

in practical applications. However, at the same time, small system preload can lead to slack in the tendon-sheath and even lead to serious problems with the wire rope coming out of the roller groove.

(2) Change in full curvature of the tendon-sheath

This group of experiments analyzes the changing characteristics of transmission efficiency by changing the full curvature size of the tendon-sheath. The preload force of the system is set to 24 N, the friction between the wire rope and the casing is dry friction, the radius of the drive roller and the load roller are both 20 mm, the relative positions of the drive motor and the load motor are adjusted, and the full curvature of the tendon-sheath is set to $\pi/2$, π and $3\pi/2$ respectively with the help of the disc, and then the torque transmission characteristic analysis experiment is conducted to measure the torque at the drive end and the load end, and the experimental results are shown in Fig. 4.21. It can be seen that the magnitude of the full curvature directly reflects the bending degree of the tendon-sheath, and the increase of the full curvature of the tendon-sheath will lead to the increase of the frictional resistance and the decrease of the transmission efficiency under the same conditions of the motion law of the system and the external load. Therefore, the actual application should optimize the design of the overall structure of the system and the alignment of the tendon-sheath to avoid excessive full curvature of the tendon-sheath, which can effectively reduce the performance requirements of the drive equipment.

(3) Changes in tendon-sheath lubrication

This set of experiments analyzes the variation characteristics of the transmission efficiency by changing the lubrication conditions between the wire rope and the casing. The preload force of the system was set to 24 N, the radius of both the driving roller and the load roller was 20 mm, and the full curvature of the sleeve was set to π rad. In the first experimental condition, the wire rope and the sleeve were in a dry

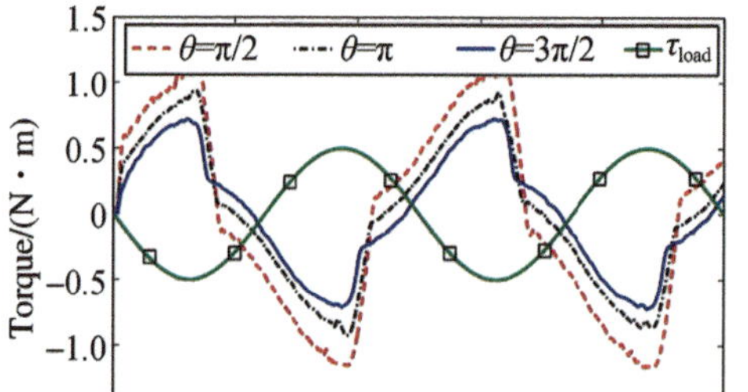

Fig. 4.21 Full curvature change [4]

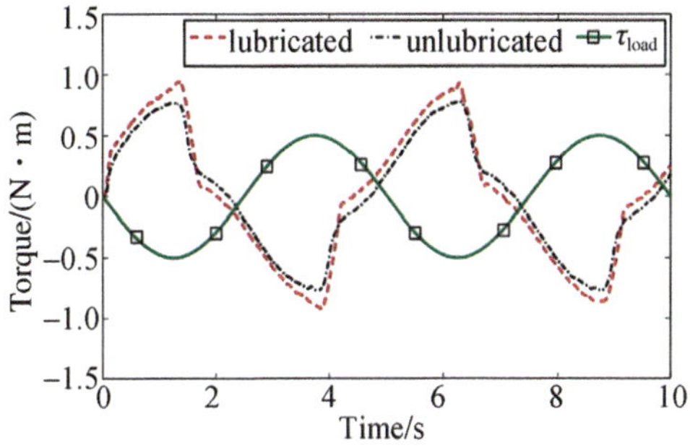

Fig. 4.22 Changes in lubrication conditions [4]

friction state without any lubrication treatment; in the second experimental condition, we attached grease (Monroe brand lubricating grease) to the wire rope and injected it into the sleeve to reduce the frictional resistance. The results are shown in Fig. 4.22. It can be seen that under the same conditions of the system motion and external load, the lubrication treatment of the tendon-sheath can help reduce the system friction and improve the transmission efficiency. Therefore, the lubrication treatment of the tendon-sheath system is necessary in practical applications.

(4) Variation of roller radius

This set of experiments was performed to analyze the variation characteristics of the transmission efficiency by changing the radii of the driving and load rollers. The preload force of the system is set to 24 N, the wire rope and the casing are in a dry friction state, and the full curvature of the tendon-sheath is set to π rad. $r_1 = r_2 = 20$ mm and $r_1 = r_2 = 30$ mm combinations of two sets of driving and load rollers are used to conduct the moment transmission characteristics analysis experiments, and the moments at the driving and load ends are measured, and the experimental results are shown in Fig. 4.23. It can be seen that, under the same motion law of the system and the same external load, the increase of the roller radius will lead to the increase of frictional resistance and the decrease of transmission efficiency. However, when the

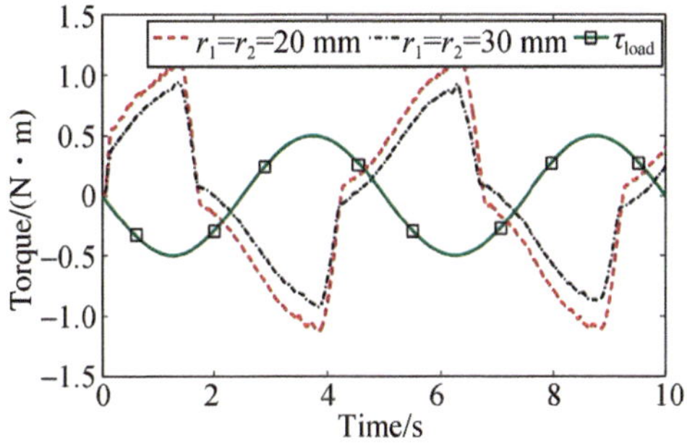

Fig. 4.23 Variation of roller radius [4]

roller radius is small, the tensile force on the wire rope in the process of movement will increase, so the performance of the tendon-sheath transmission system is required to be higher. Therefore, the roller radius needs to be optimized according to the working conditions in practical applications. Only the case of the same radius of the driving roller and the load roller is discussed here. If the radius of the two is different, the transmission ratio of the system will be changed, which will also affect the selection of the motor and reducer as well as the overall control algorithm, therefore, for the purpose of qualitative analysis, no detailed discussion is made here.

4.3 Dynamic Model of Tendon-Sheath System

The static tendon-sheath transmission model cannot explain the dynamic behavior under different moving frequency [6, 7]. The dynamic characteristics of the tendon-sheath friction shown in Fig. 4.24 displays the relationship between the friction (per unit load) and velocity at different movement frequencies. Clearly, there's noticeable friction lag in the dynamic friction force-velocity characteristics. Additionally, the dynamic friction profile shows asymmetry in different directions, indicating different friction when the system operates in positive and negative directions. It is also shown that the hysteresis loop is independent of the frequency of velocity variation.

In this context, we create a modified LuGre friction model to better capture dynamic friction characteristics. The LuGre model has been widely used to describe the dynamic friction characteristics because of its simple structure[8–10]. However, the LuGre model is not able to describe the complete tendon-sheath friction characteristics as the friction force characterized by the model is symmetric for both acceleration and deceleration phases. In current literatures, some researchers have also developed several models by modifying the LuGre friction model.

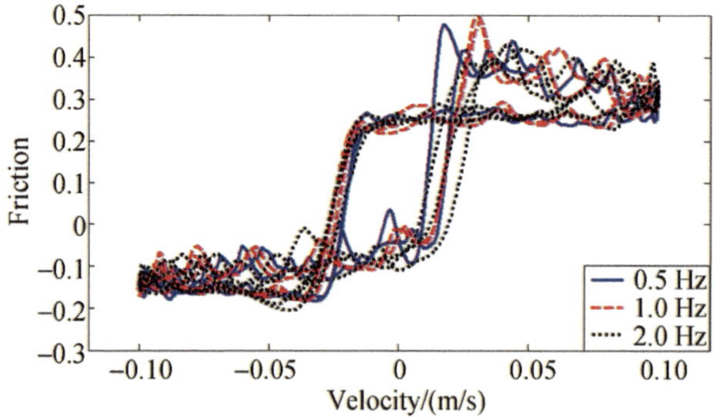

Fig. 4.24 Dynamic transmission characteristics under different frequencies of velocity variation [11]

The LuGre friction model is given by

$$\begin{cases} f = \sigma_0 z + \sigma_1 \dot{z} + \sigma_2 v \\ \dot{z} = v - \dfrac{\sigma_0 |v|}{g(\dot{x})} z \\ g(v) = F_c + (F_s - F_c)\, e^{-(v/v_s)^2} \end{cases} \tag{4.3.1}$$

where σ_0, σ_1 and σ_2 are the stiffness of the bristles, the damping coefficient and the viscous friction coefficient respectively, v is the relative velocity between contact surfaces, z is defined as the average deflection of bristles (internal state), $g(v)$ stands for an function to express the Stribeck effect, F_c and F_s indicate the Coulomb friction force and the maximum static friction.

From the steady-state transmission characteristics, the steady-state friction can be expressed as

$$f_{ss} = T_{in} - T_{out} = \left(e^{\mu\lambda\theta} - 1\right) T_{out} \tag{4.3.2}$$

where

$$f_{ss} = \begin{cases} \left(e^{-\mu\theta} - 1\right) T_{\text{out}} & v < 0 \\ \left(e^{\mu\theta} - 1\right) T_{\text{out}} & v > 0 \end{cases} \tag{4.3.3}$$

To simplify the modeling, we attribute the asymmetry of the dynamic transmission characteristics to the different steady-state friction values in different directions.

Here, a modified LuGre friction model is developed to capture the asymmetric dynamic friction characteristics. The major modification to the LuGre model is to settle the problem of the asymmetry. The method we carried out is to modify the Stribeck function $g(v)$. Firstly, the modified Stribeck function $g(v)$ is expressed as follows:

$$G(v) = gg(v)\left[F_c + (F_s - F_c)\, e^{-\left(\frac{v}{v_S}\right)^2}\right] \tag{4.3.4}$$

where

$$gg(v) = k_1 + k_2\, \text{sgn}(v) \tag{4.3.5}$$

However, the modification in Eq. (4.3.4) introduces a discontinuous phenomenon due to the use of a signum function. Thus Eq. (4.3.5) is replaced by:

$$gg(v) = k_1 + k_2\, \text{atan}(k_3 v) \tag{4.3.6}$$

It not only eliminates the discontinuous phenomenon, but also realize the smoothing transition. The second modification is to introduce an offset of friction f_0 and a relation factor of displacement $\sigma_3 x$. Finally, the proposed dynamic friction model of the tendon-sheath system can be expressed as:

$$\begin{cases} f = \sigma_0 z + \sigma_1 \dot{z} + \sigma_2 v + \sigma_3 x + f_0 \\ \dot{z} = v - \frac{\sigma_0 |v|}{g(\dot{x})} z \\ G(v) = gg(v) * \left[F_c + (F_s - F_c) e^{-\left(\frac{v}{v_S}\right)^2} \right] \end{cases} \tag{4.3.7}$$

The parameters, given in Eq. (4.3.7) can be identified using Genetic Algorithm (GA).

4.3.1 Validation of the Dynamic Transmission Model

According to the previous analysis, the hysteresis loop is independent of the frequency of the velocity variation. On this basis, we carried out a set of experiments and the results are introduced here.

In the experiment, a periodic excitation is used as the input signal. Some of the parameters in Eq. (4.3.7) are measured by experiments, and the rest are obtained using Genetic Algorithm (GA). The identified results are $K_1 = 0.045$, $K_2 = 0.205$, $K_3 = 30.2$, $\sigma_0 = -0.207$, $\sigma_1 = 0.017$, $\sigma_2 = -0.123$, $\sigma_3 = 0.241$, $f_0 = 0.088$.

Based on the identified parameters, the simulation results of the dynamic transmission model and the experimental results are shown in Figs. 4.25, 4.26 and 4.27. Figure 4.25 shows the relationship between the input displacement and the friction. Different from the static model, the dynamic model can simulate the friction during the change of transmission direction. Although the dynamic model cannot describe the frictional vibration of the tendon-sheath transmission, it has a high degree of fitting to the change curve of the friction. Figure 4.26 shows the relationship between the input velocity and the friction. Figure 4.27 depicts the time history of the friction force at an unit output load. From these figures, we can tell that the simulation results show a good agreement with the experimental results, the dynamic model can simulate the variation of friction at different transmission speeds.

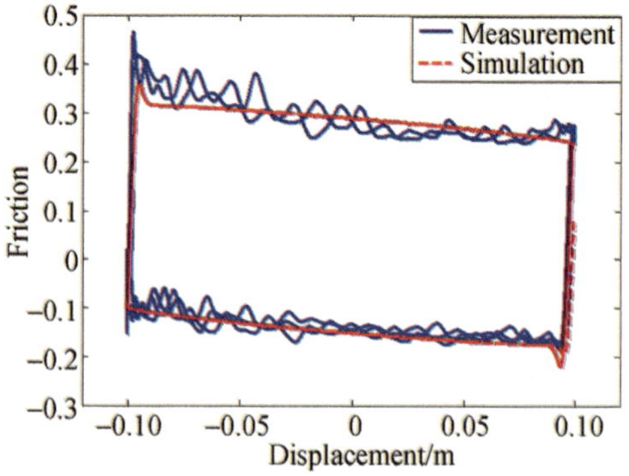

Fig. 4.25 Relationship between the input displacement and the friction [11]

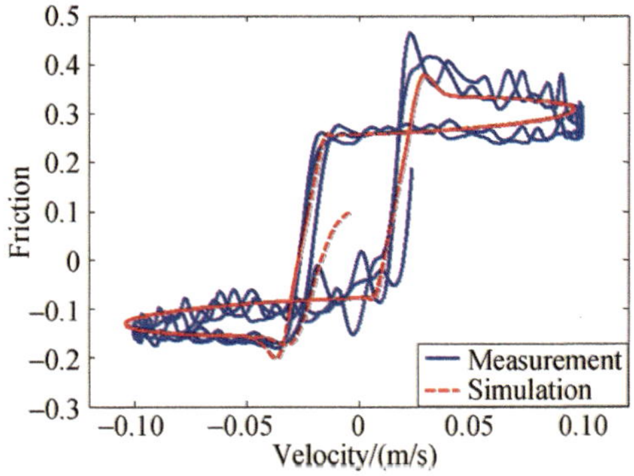

Fig. 4.26 Relationship between the input velocity and the friction [11]

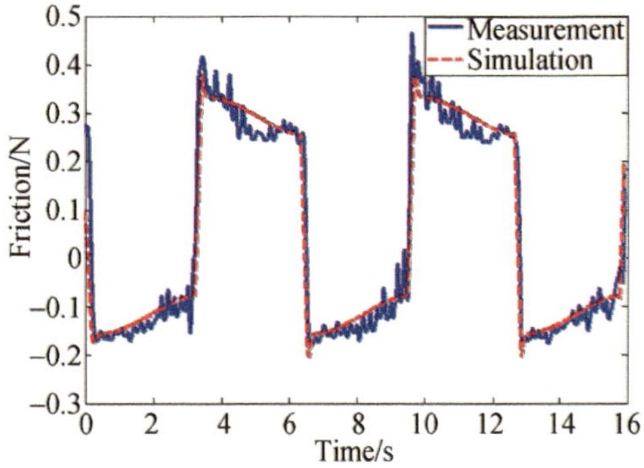

Fig. 4.27 The friction force under 1N output load [11]

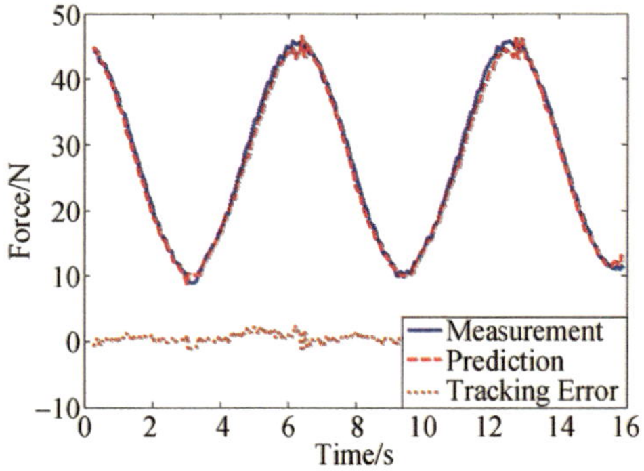

Fig. 4.28 The force at the distal end predicted by the dynamic model [11]

In addition, the dynamic transmission model can predict the force at the distal end with a good accuracy. The results are illustrated in Fig. 4.28, in which these errors are small and have lower volatility.

References

1. Kang B, Kim D, Choi H et al (2020) Learning-based fingertip force estimation for soft wearable hand robot with tendon-sheath mechanism. IEEE Robot Autom Lett 5(2):946–953
2. Liu Z, Zhang X, Cai Z et al (2021) Real-time dynamics of cable-driven continuum robots considering the cable constraint and friction effect[J]. IEEE Robot Autom Lett 6(4):6235–6242
3. Chen L, Wang X, Xu W (2013) Inverse transmission model and compensation control of a single-tendon-sheath actuator. IEEE Trans Ind Electron 61(3):1424–1433
4. Wu Q (2016) Research on gravity balanced upper limb rehabilitation exoskeleton using tendon-sheath actuator. Southeast University
5. Wu Q, Wang X, Chen L et al (2014) Transmission model and compensation control of double-tendon-sheath actuation system. IEEE Trans Ind Electron 62(3):1599–1609
6. Wu H, Yin M, Xu Z et al (2020) Transmission characteristics analysis and compensation control of double tendon-sheath driven manipulator. Sensors 20(5):1301
7. Do T, Tjahjowidodo T, Lau M et al (2014) An investigation of friction-based tendon sheath model appropriate for control purposes. Mech Syst Signal Process 42(1–2):97–114
8. Freidovich L, Robertsson A, Shiriaev A et al (2009) LuGre-model-based friction compensation. IEEE Trans Control Syst Technol 18(1):194–200
9. Do N, Ferri A, Bauchau O (2007) Efficient simulation of a dynamic system with LuGre friction
10. Marques F, Woliński, Wojtyra M et al (2021) An investigation of a novel LuGre-based friction force model. Mech Mach Theory 166:104493
11. Li J (2016) Modeling and compliance control of tendon-sheath transmission system. Southeast University
12. Chen L (2013) Modeling and control of tendon-sheath transmission system and its application in long and slender robot for searching and inspection. Southeast University

Chapter 5
Control of Tendon-Sheath Transmission Mechanism

Abstract Controlling the tendon-sheath mechanism precisely has posed challenges due to its significant nonlinearities, real-time parameter identification issues, and occasional lack of sensor feedback. Many researchers have made great efforts in compensating the inherent nonlinear behaviors and improve the control accuracy of tendon-sheath systems. Related research includes developing advanced nonlinear control methods, build more accurate transmission models, or develop sensors for tendon-sheath configuration identification, etc. This chapter outlines model-based control techniques and sheath configuration sensing for tendon-sheath systems. These approaches are simple and effective in tendon-sheath system control, and are very suitable for developing more advanced control methods on this basis.

5.1 Inverse Model Compensation with PID Control

The inverse compensation control uses the inverse transmission model to predict the system input according to the desired output [1]. As the model reflects the system transmission characteristics, the inverse compensation control offers both accuracy and speed. Additionally, it works well for robots that lack sensor feedback, like surgical robots in confined spaces or those needing frequent disinfection [2].

5.1.1 Inverse Model Compensated PID Controller Design

Take the single tendon-sheath system as an example, the inverse force and position models can be defined as Eqs. (5.1.1) and (5.1.2).

$$F_{\text{inv}}(t) = \begin{cases} \frac{F_d(t)}{Q_p}, & \dot{F}_d(t) > 0 \\ \frac{F_d(t)}{Q_r}, & \dot{F}_d(t) < 0 \\ F_{\text{inv}}\left(t^-\right), & \dot{F}_d(t) = 0 \end{cases} \tag{5.1.1}$$

X. Wang and Q. Zhang, *Tendon-Sheath Transmitted Light and Slender Robots*,
https://doi.org/10.1007/978-981-96-5581-6_5

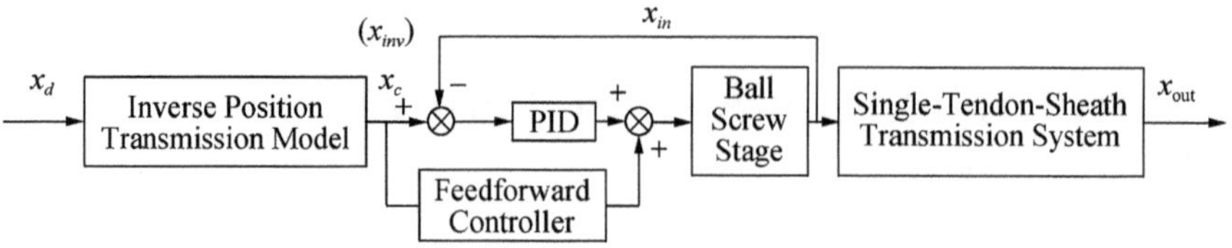

Fig. 5.1 Inverse compensation position control for tendon-sheath transmission (Reprinted from [3] in Chap. 4 with permission from IEEE)

$$x_{\text{inv}}(t) = \begin{cases} \frac{x_d(t)}{M_p}, & \dot{x}_d(t) > 0 \\ \frac{x_d(t)}{M_r}, & \dot{x}_d(t) < 0 \\ x_{\text{inv}}(t^-), & \dot{x}_d(t) = 0 \end{cases} \tag{5.1.2}$$

where F_d and x_d are the desired system outputs and F_{inv} and x_{inv} are the inverse outputs (actual system inputs). The inverse models are used to calculate the actual inputs needed at the proximal side according to the desired outputs at the distal end.

We use inverse transmission models to set up a position control system [3], as depicted in Fig. 5.1. This inverse model helps compensate for nonlinearities. The actual position command x_c that the drive stage follows comes from the position inverse model x_{inv}. When the ball screw stage moves at a low velocity, a phenomenon called stick-slip is present. This leads to significant chattering in the force signal, making force feedback less reliable. To overcome this issue, a new force control system was proposed in Fig. 5.2. Since the position and force at the output side have a linear relationship, force commands can be converted into corresponding position commands, as shown in Eq. (5.1.3).

$$x_s = \begin{cases} \frac{F_d}{kM_p}, & \text{if } \dot{F}_d > 0 \\ \frac{F_d}{kM_r}, & \text{if } \dot{F}_d < 0 \\ x_s(t^-), & \text{otherwise} \end{cases} \tag{5.1.3}$$

However, we still want the input force F_{in} to track the output force of the force inverse model. Thus, the actual position command under force control mode x_c is the sum of the stage position x_s and the position compensation x_f.

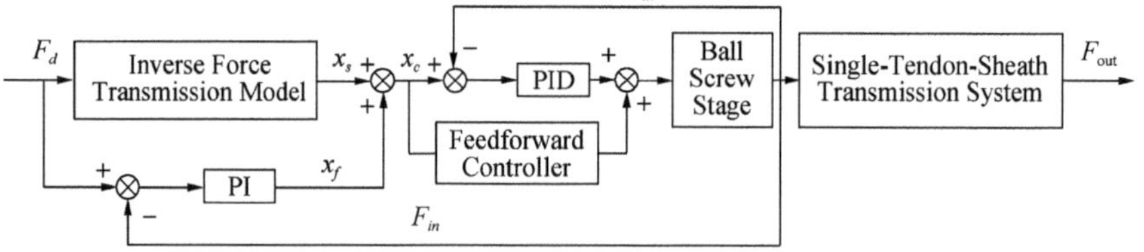

Fig. 5.2 Inverse compensation force control for the tendon-sheath transmission (Reprinted from [3] in Chap. 4 with permission from IEEE)

$$x_c = x_s + x_f \tag{5.1.4}$$

and the force modulation term is

$$x_f = K_f (F_d - F_{\text{in}}) + K_{fi} \int (F_d - F_{\text{in}})\, dt \tag{5.1.5}$$

where K_f and K_{fi} are the proportional-integral controller gains, as shown in Fig. 5.2.

5.1.2 Control Experiments and Discussion

The control results of the inverse compensation position control are shown in Fig. 5.3, it includes the tracking results with and without inverse compensation. It is found that the output position lags behind the desired trajectory due to tendon elongations when the system is without compensation. As shown in Fig. 5.3b, the position tracking performance with the compensation significantly improves, the maximal error drops from 16.8% (no compensation) to 5.3% (with compensation).

In force control mode, the system was also required to follow a sinusoid signal, the control results are presented in Fig. 5.4. It is found that without compensation, the elasticity in the tendon led to significant amplitude reduction and phase delay. The force tracking performance with the compensation is much better, with the maximal error decreasing from 66.7% (no compensation) to 25.6% (with compensation).

The results demonstrate that the inverse compensation control improves the motion and force tracking performance. Although the inverse compensation control method is developed for a single-tendon-sheath system, it can also work for other tendon-sheath systems like double tendon-sheath systems, provided the inverse model adjusts accordingly.

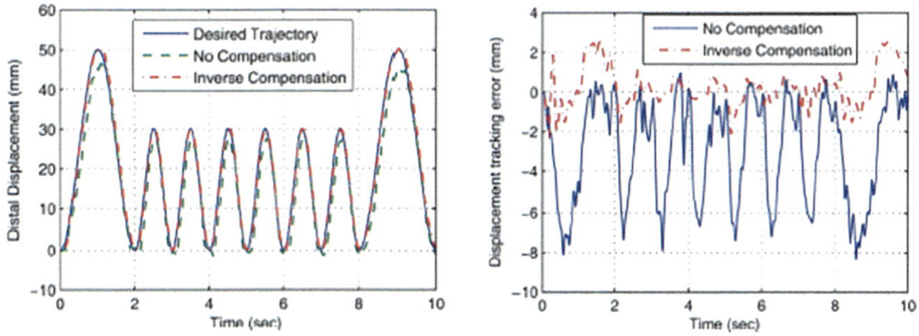

(a) Trajectory comparison (b) Control errors

Fig. 5.3 Position control results (Reprinted from [3] in Chap. 4 with permission from IEEE)

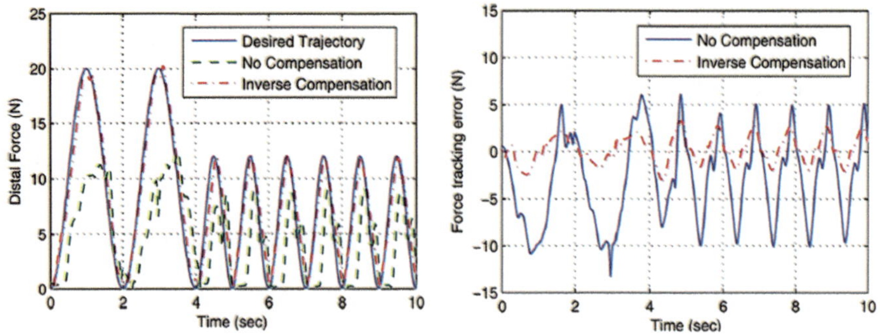

(a) Trajectory comparison (b) Control errors

Fig. 5.4 Force control results (Reprinted from [3] in Chap. 4 with permission from IEEE)

5.2 Model-Based Sliding Mode Control

To improve the accuracy, we can combine the transmission model with more advanced nonlinear control methods instead of using PID control, such as fuzzy PID and sliding mode control. Using sliding mode control, a commonly employed method for nonlinear systems, as an example, we demonstrate how to develop a model-based nonlinear position control for a tendon-sheath system [4, 5].

5.2.1 *Model-Based Sliding Mode Controller Design*

As the servo motor in the tendon-sheath actuation system runs in velocity control mode, the input signal is taken as

$$u(t) = k^{-1}\dot{x}_{in}(t) \tag{5.2.1}$$

The nonlinear behavior existing in the actuation system such as the elongation of the inner tendon and the backlash can be seen as an uncertainty system parameter, therefore the transmission characteristics can be written as

$$x_{\text{out}}(t) = x_{in}(t) + d_1(x_{in}(t)) \tag{5.2.2}$$

Taking the external disturbance d_2 into consideration, the overall system can be described as

$$\dot{x}(t) = ku(t) + d_1(x(t)) + d_2 \tag{5.2.3}$$

where $x(t)$ and $u(t)$ are the output and input signals, and $d = \dot{d}_1(x(t)) + d_2$ is the uncertainty of the system with a boundary of $|d| < D$.

The linear sliding plane will make the steady state error increase when the disturbance exists. Therefore, an integral term is added. Then the sliding plane is determined:

$$s(t) = C_1 \left[x_e(t) - \int_0^t kC_2 x_e dt \right] \tag{5.2.4}$$

where $C_1 > 0$ is the coefficient of switch, $C_2 < 0$ is the gain of state feedback. $x_e = x_d - x$ is the output error, in which x_d is the desired output.

Due to the uncertainty and disturbance of the system, the sliding mode control law composed of equivalent control and switching control is adopted as follow

$$u = k^{-1}\dot{x}_d - C_2 x_e + k^{-1} D\,\mathrm{sgn}(s) \tag{5.2.5}$$

As the upper bound of the overall uncertainty D is usually hard to be obtained, an adaptive method is proposed. Suppose that $\hat{D}$ is the estimation of the upper bound D, then the adaptive law is defined as

$$\dot{\hat{D}} = \frac{1}{\alpha} |C_1 s| \tag{5.2.6}$$

where α is the gain of adaptive term.

Hence, the robust adaptive sliding mode control law is expressed as

$$u = k^{-1}\dot{x}_d - C_2 x_e + k^{-1} \hat{D}\,\mathrm{sgn}(s) \tag{5.2.7}$$

To analyze the stability of the control method, Lyapunov function V is defined as

$$V = \frac{1}{2}s^2 + \frac{1}{2}\alpha \tilde{D}^2 \tag{5.2.8}$$

in which $\tilde{D} = \hat{D} - D$.

The derivative of s can be calculated as follow

$$\begin{aligned} \dot{s} &= C_1 \dot{x}_e - C_1 C_2 k x_e \\ &= C_1 (\dot{x}_d - ku - d) - C_1 C_2 k x_e \\ &= C_1 \left(\dot{x}_d - \left(\dot{x}_d - C_2 k x_e + \hat{D}\,\mathrm{sgn}(s) \right) - d \right) - C_1 C_2 k x_e \\ &= C_1 (-d - \hat{D}\,\mathrm{sgn}(s)) \end{aligned} \tag{5.2.9}$$

Additionally

$$\alpha \tilde{D} \dot{\hat{D}} = \alpha \tilde{D} \frac{1}{\alpha} |C_1 s| = C_1 \tilde{D} |s| \tag{5.2.10}$$

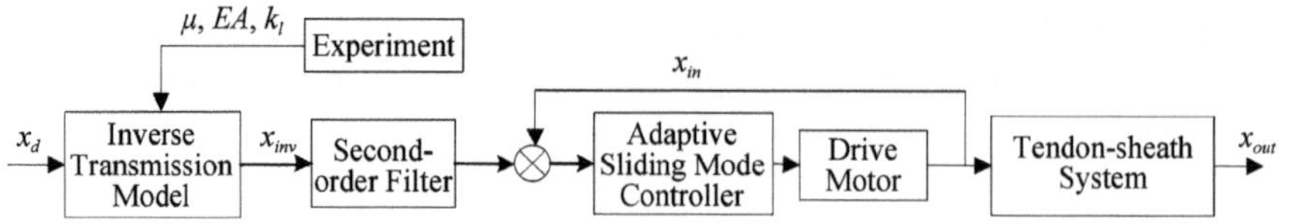

Fig. 5.5 The model-based sliding model control for tendon-sheath system

Then the derivative of V can be expressed as

$$\begin{aligned}\dot{V} &= s\dot{s} + \alpha\tilde{D}\dot{\tilde{D}} \\ &= C_1(-ds - \hat{D}|s|) + C_1\tilde{D}|s| \\ &= C_1(-ds - D|s|) \le 0\end{aligned} \tag{5.2.11}$$

From Eqs. (5.2.8) and (5.2.11), it can be seen that V is positive definite Whereas $\dot{V}$ is semi-negative definite, which proves that this control law is stable. Combining with Eq. (5.1.2), we can set up the control method as Fig. 5.5 shows, in which the input position x_{in} is fed back by the sensors at the proximal end.

5.2.2 Control Experiments and Discussion

The parameters of the adaptive sliding control method in the experiments are as follows: $k^{-1} = 0.0365$, $c_1 = 0.12$, $c_2 = -0.22$, $\alpha = 1.25$, the control results are shown in Fig. 5.6.

Comparing to the uncompensated control method that lacks the inverse model compensation, the model-based sliding model control greatly improves the performance of position tracking: the maximum absolute tracking error decreases from 4.7 mm to around 1.9 mm, and the maximum full-scale error is 3.5%.

Note that the inverse compensation control is an open-loop control, once the sheath shape changes, the system parameters have to be updated. This method is particularly useful for those tendon-sheath actuated robots that have to work in confined space, such as NOTES robots and rescue robots. Once they reach the working site, the robot shape does not change or changes slightly, and neither does the global sheath shape. However, if the sheath shape changes frequently during the working process, this method is not applicable. To address this, some model-based close-loop control methods were developed.

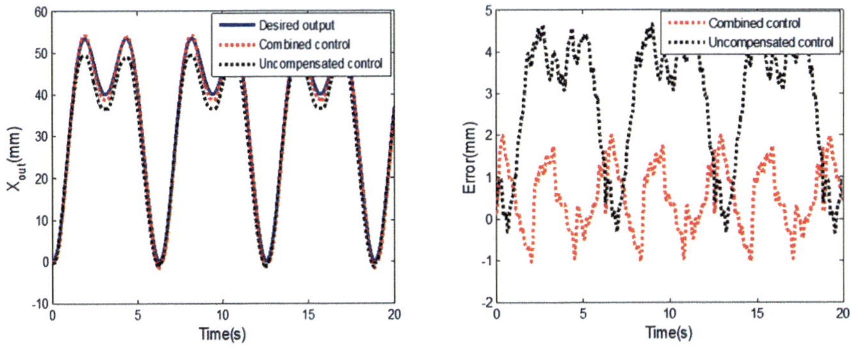

(a) The position tracking performance (b) The position tracking errors

Fig. 5.6 The experimental results of the model-based sliding model control (Reprinted from [4] with permission from IEEE)

5.3 Neural Network-Based Sliding Mode Control

The sliding mode control shows improved accuracy in the control of tendon-sheath transmission, however, the inherent chattering problem of conventional SMC scheme may lead to unpredictable instabilities. In this section, we introduce a neural network-based sliding mode control (NNSMC) for tendon-sheath mechanism [6]. The developed NNSMC scheme is capable of estimating the nonlinear system dynamics with disturbance and improve the position-control accuracy of tendon-sheath transmission. Now, we still take the single tendon-sheath transmission as an example to introduce the establishment and application process of the NNSMC method.

5.3.1 NNSMC Position-Controller Design

Consider the case of a single tendon-sheath with an end load, we can establish the system dynamic model as a mass-spring-damper system [7].

$$M_g\ddot{x}(t) + B_g\dot{x}(t) + K_g x(t) + P_g(t) = F_g(t) \tag{5.3.1}$$

where x is the measured displacement of gripper; t represents the time variable; parameters M_g, B_g and K_g denote the mass, damping coefficient and stiffness of the end load; $P_g(t)$ is the lumped effects of external disturbance and uncertainty and $F_g(t)$ is the output force of the tendon-sheath mechanism.

According to the force transmission model of single tendon-sheath transmission, the $F_g(t)$ can be described as the following equation

$$F_g(t) = F_m(t) \exp\left[\lambda\mu \int_0^L \kappa(\delta, t) d\delta\right] \tag{5.3.2}$$

where L denotes the length of outer sheath; δ is the position of infinitesimal tendon-sheath element; $F_{\mathrm{m}}(t)$ represents the pulling force acting on the tendon of actuation side; λ denotes the moving direction of tendon; μ represents the Coulomb friction coefficient; $\kappa(\delta, t)$ and $\theta(s, t)$ are the curvature and arc-angle of infinitesimal tendon-sheath element.

The relation between the output torque of driving motor and the pulling force acting on the tendon of actuation side can be expressed as follow

$$F_m(t) = \frac{\tau(t)}{r} \tag{5.3.3}$$

where $\tau(\sigma, t)$ represents the output torque and r denotes the radius of pulley assembled on the motor.

Therefore, by substituting Eqs. (5.3.2) and (5.3.3) into Eq. (5.3.1), the dynamics model can be refined as follows

$$M\ddot{x}(t) + B\dot{x}(t) + Kx(t) + P(t) = \tau(t) \tag{5.3.4}$$

in which,

$$M = \frac{M_g r}{\exp\left[\lambda\mu \int_0^L \kappa(\delta, t) d\delta\right]}$$

$$B = \frac{B_g r}{\exp\left[\lambda\mu \int_0^L \kappa(\delta, t) d\delta\right]}$$

$$K = \frac{K_g r}{\exp\left[\lambda\mu \int_0^L \kappa(\delta, t) d\delta\right]}$$

$$P(t) = \frac{P_g(t) r}{\exp\left[\lambda\mu \int_0^L \kappa(\delta, t) d\delta\right]}$$

where M, B, K and $P(t)$ represent the refined mass, refined damping coefficient, refined stiffness and refined lumped effects of uncertainty.

According to Eq. (5.3.4), the system uncertainty $P(t)$ can be expressed as

$$P(t) = \tau(t) - M\ddot{x}(t) - B\dot{x}(t) - Kx(t) \tag{5.3.5}$$

However, the direct use of Eq. (5.3.5) will lead to algebraic loops, and it is unsuitable for practical application. Thus, the perturbation estimation technique is employed in this article to estimate the disturbance [8].

$$\hat{P}(t) = \tau(t - T) - M\ddot{x}(t) - B\dot{x}(t) - Kx(t) \tag{5.3.6}$$

where $\hat{P}(t)$ denotes the estimated value of system uncertainty; T is the sampling time interval of control system and $\tau(t - T)$ represents the driving torque in the previous time step. The sampling time is selected small enough to reduce the estimation error of lumped uncertainty $\tilde{P}(t)$, which can be described as the following equation

$$\tilde{P}(t) = P(t) - \hat{P}(t) \tag{5.3.7}$$

Assume that the boundary of estimation error can be given as follow

$$|\tilde{P}(t)| < \eta \tag{5.3.8}$$

where η is a positive constant denoting the upper boundary of estimation error.

By substituting Eq. (5.3.7) into Eq. (5.3.5), the dynamics model can be rewritten as

$$\ddot{x}(t) = M^{-1}[\tau(t) - B\dot{x}(t) - Kx(t) - \hat{P}(t) - \tilde{P}(t)] \tag{5.3.9}$$

In order to derive the NNSMC-based position control algorithm, the position-tracking error, velocity tracking error and acceleration-tracking error of the system output are defined, respectively, as follows

$$\begin{aligned} e(t) &= x_d(t) - x(t) \\ \dot{e}(t) &= \dot{x}_d(t) - \dot{x}(t) \\ \ddot{e}(t) &= \ddot{x}_d(t) - \ddot{x}(t) \end{aligned} \tag{5.3.10}$$

where $x_d(t)$, $\dot{x}_d(t)$ and $\ddot{x}_d(t)$ represent the desired position, velocity and acceleration, respectively.

An auxiliary parameter $\dot{x}_r(t)$ is defined as follows

$$\dot{x}_r(t) = \dot{x}_d(t) + \Lambda e(t) \tag{5.3.11}$$

where Λ represents a designed positive constant.

Then, the sliding switching function for sliding surface can be defined as follow

$$S(t) = \dot{e}(t) + \Lambda e(t) = \dot{x}_r(t) - \dot{x}(t) \tag{5.3.12}$$

The derivative of the sliding variable with respect to t can be expressed as

$$\dot{S}(t) = \ddot{e}(t) + \Lambda\dot{e}(t) = \ddot{x}_r(t) - \ddot{x}(t) \tag{5.3.13}$$

The radial basis functional network (RBFN) has been effectively used for nonlinear function approximation and pattern recognition [9]. A typical three-layer RBFN is used in this method. The Gaussian function is chosen as the receptive field unit of RBFN. Assuming that there are M_K hidden nodes in the RBFN, and $\phi(\xi) = (\phi_1(\xi), \phi_2(\xi), \ldots, \phi_M(\xi))^{\mathrm{T}}$ is chosen as the selected Gaussian function. The output value can be calculated using the weighted sum approach expressed as the following two equations.

$$f_i = \sum_{j=1}^{M_K} c_{ij}\phi_j, \quad i = 1, 2, \ldots, N_K \tag{5.3.14}$$

$$\phi_j(\xi) = \exp\left[-\frac{(\xi - m_j)^T(\xi - m_j)}{2b_j^2}\right] \tag{5.3.15}$$

where f_i represents the value of the ith output node. The numbers of output nodes are expressed as N_K. The input vector of RBFN is denoted as ξ.$\phi(\xi)$ represents the Gaussian function value for the jth neural net in the hidden layer. c_{ij} denotes the weight connecting the jth hidden node to the ith output node. m_j denotes the center vector of the receptive field. b_j denotes the standard deviation of the jth neuron unit.

From Eqs. (5.3.4), (5.3.12) and (5.3.13), we can get

$$M\dot{S}(t) = -BS(t) - \tau(t) - P(t) + f \tag{5.3.16}$$

where the nonlinear robot function can be expressed as

$$f = M\ddot{x}_r(t) + B\dot{x}_r(t) + Kx(t) \tag{5.3.17}$$

According to the RBFN with finite hidden nodes, the estimation of the nonlinear robot function can be expressed as the following continuous equations

$$f = W^{*T}\phi(\xi) + \varepsilon(\xi) \tag{5.3.18}$$

$$\|\varepsilon(\xi)\| \leqslant \varepsilon_N \tag{5.3.19}$$

where W^* represents the ideal weight of approximation; $\xi = [x, \dot{x}, \dot{x}_r, \ddot{x}_r]$ denotes the input vector; $\varepsilon(\xi)$ is the RBFN approximation error; ε_N represents the upper bound of RBFN approximation error.

Selecting the control law of the proposed NNSMC strategy as the following equations

$$\mu = \hat{W}^T\phi(\xi) + \beta S(t) + \hat{P}(t) + \alpha\,\mathrm{sat}\{S(t)\} \tag{5.3.20}$$

$$\mathrm{sat}\{S(t)\} = \begin{cases} 1 & \text{if } |S(t)| > \sigma \\ \frac{(1-\rho)S(t)}{\sigma} & \text{if } |S(t)| \leqslant \sigma \end{cases} \tag{5.3.21}$$

$$\dot{\hat{W}} = H\phi\left(x, \dot{x}, \dot{x}_r, \ddot{x}_r\right) S(t) \tag{5.3.22}$$

where $\hat{W}$ is the estimation value of ideal weight; α and β represent two positive gains; $\mathrm{sat}(s)$ denotes a saturation function with a constant dead-zone compensation ρ; the boundary layer thickness of saturation function is denoted as σ and H denotes a positive constant. The perturbation estimation and the neural network can be implemented via the control module of the RTW environment, which helps to reduce the computation burden of the developed NNSMC strategy.

The Lyapunov stability criteria is employed to verify the closed-loop stability of the proposed NNSMC control strategy. A positive-definite Lyapunov function candidate is adopted as follow

$$V = \frac{1}{2}\left\{MS^2(t) + H^{-1}\tilde{W}^T\tilde{W}\right\} \tag{5.3.23}$$

where $\tilde{W} = W^* - \hat{W}$ represents the error between the idea weight and estimation weight of approximation.

Differentiating both sides of Eq. (5.3.23) and utilizing Eqs. (5.3.9) and (5.3.13), we can get

$$\begin{aligned} \dot{V} &= MS(t)\dot{S}(t) - H^{-1}\tilde{W}^T\dot{\hat{W}} \\ &= MS(t)\left\{\ddot{x}_r(t) - M^{-1}[\tau(t) - B\dot{x}(t) - Kx(t) - \hat{P}(t) - \tilde{P}(t)]\right\} - H^{-1}\tilde{W}^T\dot{\hat{W}} \\ &= MS(t)\left\{\ddot{x}_r(t) - M^{-1}\left[\hat{W}^T\phi(\xi) + \beta S + \alpha\,\mathrm{sat}(S) - B\dot{x}(t) - Kx(t) - \tilde{P}(t)\right]\right\} - H^{-1}\tilde{W}^T\dot{\hat{W}} \end{aligned} \tag{5.3.24}$$

Combining equations Eqs. (5.3.17) and (5.3.18) into Eq. (5.3.24), we can obtain

$$\begin{aligned} \dot{V} &= MS(t)\left\{M^{-1}[f - B\dot{x}_r(t) - Kx(t)] - M^{-1}\left[\hat{W}^T\phi(\xi) + \beta S(t) + \alpha\,\mathrm{sat}(S) - B\dot{x}(t) - Kx(t) - \tilde{P}(t)\right]\right\} - H^{-1}\tilde{W}^T\dot{\hat{W}} \\ &= MS(t)\left\{M^{-1}\left[W^{*T}\phi(\xi) + \varepsilon(\xi) - B\dot{x}_r(t) - Kx(t)\right] - M^{-1}\left[\hat{W}^T\phi(\xi) + \beta S(t) + \alpha\,\mathrm{sat}(S) - B\dot{x}(t) - Kx(t) - \tilde{P}(t)\right]\right\} \\ &\quad - H^{-1}\tilde{W}^T\dot{\hat{W}} \\ &= S(t)\left\{W^{*T}\phi(\xi) - \hat{W}^T\phi(\xi) + \varepsilon(\xi) - B\left[\dot{x}_r(t) - \dot{x}(t)\right] - \beta S(t) - \alpha\,\mathrm{sat}(S) + \tilde{P}(t)\right\} - H^{-1}\tilde{W}^T\dot{\hat{W}} \\ &= S(t)\{\varepsilon(\xi) - BS(t) - \beta S(t) - \alpha\,\mathrm{sat}(S) + \tilde{P}(t)\} + S(t)\tilde{W}^T\phi(\xi) - H^{-1}\tilde{W}^T\dot{\hat{W}} \end{aligned} \tag{5.3.25}$$

By substituting Eq. (5.3.22) into Eq. (5.3.25), we can get that

$$\begin{aligned} \dot{V} &= S(t)\{\varepsilon(\xi) - BS(t) - \beta S(t) - \alpha\,\mathrm{sat}(S) + \tilde{P}(t)\} \\ &= S(t)\{\varepsilon(\xi) + \tilde{P}(t) - \alpha\,\mathrm{sat}(S)\} - (B+\beta)S^2(t) \\ &\leqslant |S(t)|\{|\varepsilon(\xi)| + |\tilde{P}(t)| - \alpha\,\mathrm{sat}(|S|)\} - (B+\beta)S^2(t) \\ &\leqslant |S(t)|\{|\varepsilon(\xi)| + |\tilde{P}(t)| - \alpha\rho\} - (B+\beta)S^2(t) \end{aligned} \tag{5.3.26}$$

The refined damping coefficient, that is, B, has been estimated through parameters identification experiment. Assuming that the difference between the refined damping coefficient and identified damping coefficient is bounded, we can get

$$|B - \hat{B}| \leqslant B_N \tag{5.3.27}$$

where B_N is the upper bound. Thus

$$|B| \leqslant B_N + |\hat{B}| \tag{5.3.28}$$

If the positive gains α and β are selected as follows

$$\begin{cases} \alpha \geqslant \frac{\eta+\varepsilon_N}{\rho} \geqslant \frac{|\alpha(\xi)|+|\bar{P}(t)|}{\rho} \\ \beta \geqslant B_N + |\hat{B}| \end{cases} \tag{5.3.29}$$

By substituting Eq. (5.3.29) into Eq. (5.3.26), we can get that

$$\dot{V} \leqslant 0 \tag{5.3.30}$$

The result indicates that V is positive definite, while $\dot{V}$ is a negative definite. Therefore, the NNSMC control law for tendon-sheath mechanism is stable in Lyapunov sense. The position-tracking error gradually converges to 0 and approaches the sliding surface (i.e. $s = 0$) in finite time.

5.3.2 Control Experiments and Discussion

To validate the effectiveness of the NNSMC control algorithm, an elastic gripper with linear force-displacement relation was used as the end load, the overall block diagram of the proposed NNSMC strategy is depicted in Fig. 5.7.

In the experiment, a sinusoidal trajectory-tracking test was conducted. The desired displacement was a sinusoidal trajectory with time-varying amplitude and frequency. The first 12 s of predefined trajectory has an amplitude of 12 mm and a frequency of 0.25 Hz. Then, the next 6 s has an amplitude of 10 mm and a frequency of 0.5 Hz. In the last 2 s, the amplitude and frequency turn to 8 mm and 1 Hz. The parameters

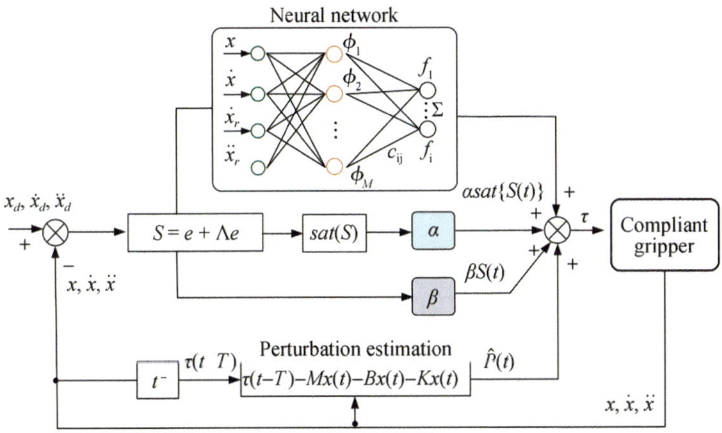

Fig. 5.7 Overall block diagram of the developed neural network based sliding mode control strategy

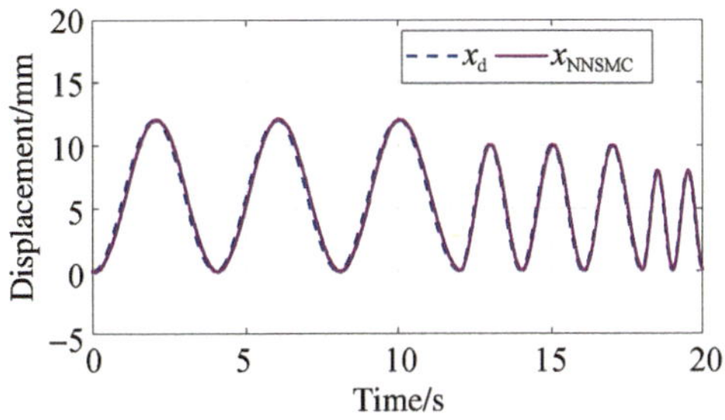

Fig. 5.8 Trajectory-tracking result of conventional NNSMC controller

of the proposed NNSMC method were empirically adjusted by intensive experiments, which were set to the optimal values as follows: $\Lambda = 1.65$, $T = 0.0001$, $\rho = 0.32$, $\sigma = 1.26$, $\alpha = 0.0015$ and $\beta = 0.0026$.

The experimental results were shown in Fig. 5.8. It can be clearly found that the gripper can follow the desired trajectory with high accuracy. The NNSMC control scheme produces maximum error, mean absolute error and root mean squared error (RMSE) of 1.16, 0.47 and 0.53 mm, which equal to 9.66, 3.91 and 4.41% of the motion range, respectively. These results demonstrated that the NNSMC control is fast and precise, and is effective for tendon-sheath mechanism with external uncertainties to achieve high accuracy.

It should be pointed out that the dynamic model changes with the variation of the tendon-sheath configuration. Therefore, the proposed control strategy is not suitable for the application with time-varying tendon-sheath configuration. Thus, our future works will be devoted to developing an online identification method to overcome this limitation.

References

1. Li Z, Shan J, Gabbert U (2020) A direct inverse model for hysteresis compensation. IEEE Trans Ind Electron 68(5):4173–4181
2. Shao Z, Wu Q, Chen B et al (2019) Modeling and inverse control of a compliant single-tendon-sheath artificial tendon actuator with bending angle compensation. Mechatronics 63:102262
3. Li J (2016) Modeling and compliance control of tendon-sheath transmission system. Southeast University
4. Wang X, Zhang Q, Shen X et al (2018) Noncollocated position control of tendon-sheath actuated slender manipulator. IEEE Trans Control Syst Technol 28(2):688–696
5. Zhang Q, Shen D, Tian M et al (2022) Model-based force control of a tendon-sheath actuated slender gripper without output feedback. J Intell Robot Syst 106(4):79
6. Wu Q, Wang X, Chen B et al (2019) Neural network-based sliding-mode control of a tendon sheath-actuated compliant rescue manipulator. Proc Inst Mech Eng Part I: J Syst Control Eng 233(8):1055–1066
7. Kazerooni H, Houpt P, Sheridan T (1986) Robust compliant motion for manipulators, part II: design method. IEEE J Robot Autom 2(2):93–105
8. Elmali H, Olgac N (1996) Implementation of sliding mode control with perturbation estimation (SMCPE). IEEE Trans Control Syst Technol 4(1):79–85
9. Li T, Duan S, Liu J et al (2015) A spintronic memristor-based neural network with radial basis function for robotic manipulator control implementation. IEEE Trans Syst, Man, Cybern: Syst 46(4):582–588

Chapter 6
Mechanism of Tendon-Sheath Transmitted Slender Robots

Abstract Slender robots, like trunk robots, snake robots, are popular for tasks such as surgery, detection and rescue missions. They have multiple degrees of freedom and flexible bodies, allowing them to operate in narrow and complex spaces. However, the drive system and transmission mechanism remains a challenge in the design of slender body. Compared to other actuation mechanisms, the tendon-sheath actuation system is flexible to adapt to the complex environment. Additionally, it allows for a separate design for the robot's actuation and drive mechanisms, such that enabling remote operations. In this chapter, we introduce various prototypes of slender robots actuated by the tendon-sheath system. We delve into the design of this mechanism and explore how to control such flexible and slender robot systems. This chapter aims to offer examples and insights for those working on slender robot development, addressing some common design and application challenges.

6.1 Macro-Micro Configurations of Slender Robots

Slender robots possess unique advantages in various domains, particularly in search and rescue operations within disaster-stricken areas, as well as in detecting small spaces within equipment and other relevant fields [1, 2]. To address these requirements, the concept of a "slender robot with tendon-sheath transmission" has been introduced. Figure 6.1 illustrates the fundamental configuration of this robot. The drive unit, situated in the robot's base, plays a pivotal role in controlling the motion of the distal joint and the end effector. This control is achieved through the utilization of tendon-sheath transmission, enabling the robot to effectively execute search and rescue missions or carry out detection tasks.

The slender robot usually has a flexible or rigid casing that wraps the tendon-sheath mechanism, serving as the transmission pathway for the end effector. We categorize the slender robot into two types: controllable deformation and slender-free. In the case of a controllable deformation configuration, the structure resembles that of continuous robots, such as robotic snakes and elephant trunk robots [3, 4]. On the other hand, when the body shape is slender and free, it represents a novel category of robots referred to as soft and slender robots. Notably, the latter body

X. Wang and Q. Zhang, *Tendon-Sheath Transmitted Light and Slender Robots*,
https://doi.org/10.1007/978-981-96-5581-6_6

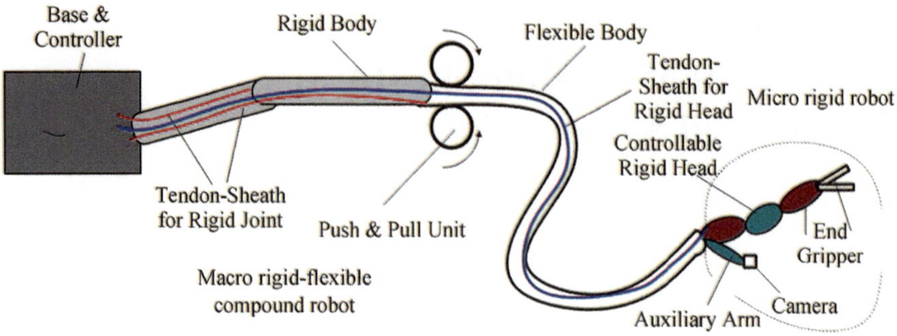

Fig. 6.1 Slender robot with tendon-sheath transmission

deformation is not restricted and exhibits greater flexibility, making it highly suitable for disaster search and rescue missions and for detecting small internal spaces within the equipment. It is worth mentioning that endoscopic robots can be classified as soft and slender robots, but the current endoscopic end effectors lack deformation controllability. Hence, our research is focused on overcoming this limitation.

The slender robot we discussed is a prototypical macro-micro manipulator (MMM) structure [5]. In this configuration, the slender body can be classified as the macromanipulator, while the multi-joint manipulator at the end can be regarded as the micromanipulator. Conventionally, the macromanipulator in MMM can either be rigid manipulators or flexible manipulators while the micromanipulator is typically a rigid manipulator. However, with the use of the tendon-sheath mechanism, the macromanipulator become slender rigid-soft hybrid manipulator, whereas the micromanipulator can be controllable multiple rigid joints or flexible manipulators.

In recent years the MMM research has been extended to the dynamic control of flexible manipulators. A rigid small robot is attached at the end of the flexible manipulator, forming the so-called macro-flexible and micro-rigid manipulator, and its fine motion is applied to compensate for the positioning or tracking error of the flexible macro manipulator.

The development of the MMM originated from the need for better accuracy and dynamic performance. To the best knowledge of the authors, the MMM was first mentioned explicitly in the study of Sharon and Hardt11 in 1984 [6]. Khatib, in his work, analyzed the control and effective inertial characteristics of MMM [7]. In [8], a general control framework for macro/mini manipulator to improve the force and compliant motion control performance is introduced. In [9], hybrid position/force controllers are used on a mechanism which is the combination of a flexible macro and a rigid micro mechanism. Henceforth, the MMM became a field of study for improving positional and force accuracy as well as dexterous manipulation.

In terms of applications, many studies are conducted to enhance the performance of systems by means of accuracy and force control. An MMM welding robot based on the vision navigation is designed in [10], which is used for seam tracking. The control is achieved by the task space division method in which motion is planned for the macro manipulator, while the micro manipulator is used to compensate the tracking errors in real time.

The experimental system in this study comprises 7-DoF Mitsubishi PA-IO as the macro manipulator and 1-DoF voice coil as the mini manipulator. A force/position controller is devised so that it could reduce the impact when the manipulator makes contact with the workpiece. In this case, the micro manipulator is used to regulate the contact forces during the motion of the macro manipulator. MMM control architecture is developed in [11] to enhance the accuracy of robotic machining. An industrial robot, KUKA model KR125, is used as the macro manipulator, while an external piezo-actuated compensation mechanism is used as the micro manipulator. This integrated system is reported to offer significantly higher bandwidth control of the relative position between the tool and the workpiece. The experimental results show that the proposed approach offers higher accuracy, with up to eight times improvement in the milling operation.

Another approach that utilizes MMM is shown to enhance the acceleration capability of the total system by adding a lightweight micro manipulator to a conventional machine. Trumpf company devised a solution in their planar laser cutting machine called Trumatic 7000 to increase the acceleration of the machine by adding additional axes alongside the X- and Y-axes, namely, making the configuration of the machine XY+xy. Prima Power Company, from Italy, designed Syncrono Series planar laser cutting machines with two additional axes. Syncrono has the capability of reaching a combined acceleration of 6g. In this system, the conventional machine axes (X, Y) which can be named as the macro manipulator axes are used to position laser head for macro displacements, and the additional two axes (u and v), which can be named as the micro manipulator axes, are used to move the laser head in a relatively smaller work space.

The MMM that is used in this study is a simplified model of the kinematically redundant machine in [12], which is designed for the planar laser cutting operation. Similar to the applications mentioned above, the MMM concept is implemented by adding a micro mechanism at the distal end of a conventional Cartesian manipulator (an x-y table). However, the parallel mechanism used for the micro manipulation and redundancy resolution algorithms used in this study is different from the previous examples in the industry. The agility of the relatively smaller sized parallel mechanism with lower inertia with respect to the macro mechanism is used to enhance the total acceleration capability of the system while the work space dimensions are preserved. In this previous study, a trajectory generation scheme with an off-line task division method and an online compensation to generate the micro mechanisms motion trajectory is proposed and employed on an industrial laser cutting machine. The main novelty of the study presented in this article is that it limits the maximum acceleration of the macro manipulator within the desired acceleration bounds during the operation without the need for an iterative procedure as presented in [13]. The proposed methodology also results in the minimization of the total actuation effort.

According to the macro-micro manipulator concept, the slender robot is categorized into two main components: the end microrobot and the macro slender robot. Utilizing existing theories and methodologies, comprehensive theoretical and application-oriented research can be conducted on the tendon-sheath driven slender robot. In this book, continuum robots serve as the foundational point of investigation for the exploration of "slender robots". These continuum robots including snake-shaped robots and elephant-trunk robots, are designed with inspiration from biological structures. As engineering applications progress, many slender endoscopic robots and interventional puncture robots have also emerged.

6.2 Morphological Character of Tendon-Sheath Transmitted Slender Continuum Robots

The morphology of continuum robots is the basic starting point for the study of "soft elongated robots" in this book. According to their bionic objects, continuum robots include snake robots, elephant trunk robots; with the development of engineering applications, endoscopic elongated robots, interventional puncture elongated robots, and there are many kinds of robots. This chapter focuses on the basic working principle of tendon-sheath-driven continuum macro elephant trunk robots and their institutional design and kinematic characteristics.

To further refine the elongated body structure of the elongated robots, the possible elongated body structures can be illustrated in Fig. 6.2. As can be seen, the macro-robots here can be either flexible robotic arms (Fig. 6.2a–e) or soft-bodied serpentine or rope-like (endoscopic) structures (Fig. 6.2d, f). Among them, the soft and slender robots, also known as continuum robots, include robotic snakes, robotic elephant trunks, and endoscopic robots. Initially they each modelled organisms of different structures, but they are characterised by continuous deformation and are therefore often classified as continuous robots.

There are various methods to design continuous robots. Currently, these robots have elongated body parts and are designed to be controllable [14, 15]. As most continuous robots are used for minimally invasive surgeries and have predictable shapes and lengths, they may not always meet the requirements for unpredictable disaster scenarios. In such cases, snake-like or endoscopic robots are more suitable due to their flexibility and adaptability. How to make the slender robots have controllable body configuration and flexible and accurate end manipulation depends on the driving of tendon-sheath system, which attracts us to carry out the theoretical research on the tendon-sheath transmission and form this book.

Through the macro-micro robot division and research methods, we can carry out the study of elongated robots by drawing on macro robots such as rigid joint robots, flexible arm robots, and continuum robots; and then combine them with micro-rigid robots to carry out the mechanics, dynamics and control of soft elongated robots, and to achieve intended function. Later sections in this chapter will present the mechanics kinematics of flexible articulated arm macro robotics for elongated robots, tendon-sheath-driven continuum macro robotics, and the mechanics and kinematics

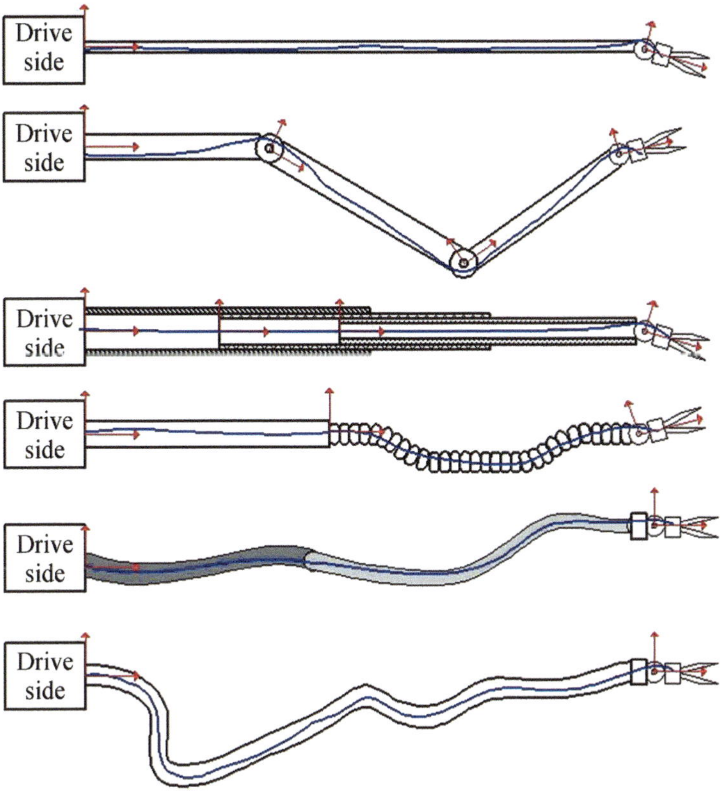

Fig. 6.2 Slender body structure

of tendon-sheath-driven micro-articulated arm robotics, and tendon-sheath-driven micro-rigid mechanism robotics, respectively.

6.3 Mechanism of Tendon-Sheath Transmitted Macro Slender Link Series

6.3.1 Modular Joint Design of the Slender Link Series

The working environment of the slender robot is unpredictable, so it need to be versatile for different tasks. In order to achieve high mobility of the slender robot, we can adopt the design concept of modular joints to construct the series link joints to form the slender body or end manipulator. The modular robot system has various components like joint modules, end-effectors, and control parts, each of these components offers unique functions and dimensions, allowing for specific assembly structures.

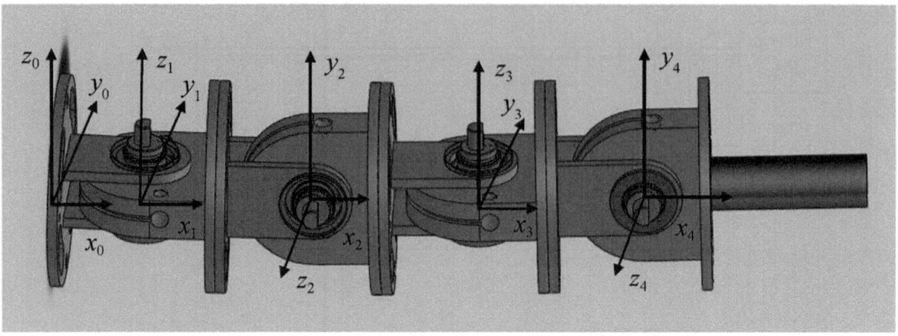

Fig. 6.3 Series link joints for slender robot [17]

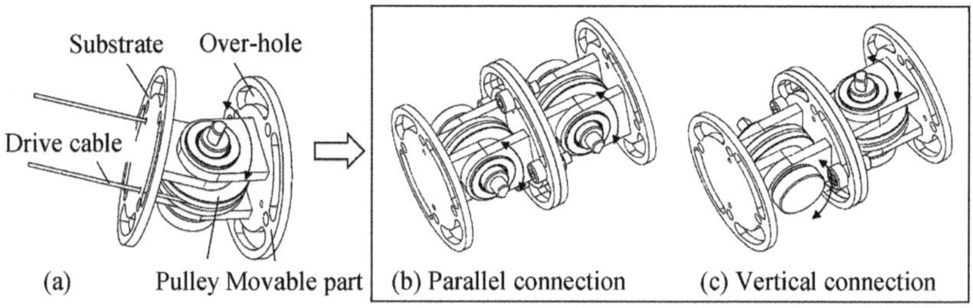

Fig. 6.4 Structure diagram of deflection joint module [17]

Notably, the modular design boasts advantages such as easy reconfiguration, flexibility, straightforward assembly, and maintenance [16]. Figure 6.3 shows a simple slender link series structure formed by a series of modular joints., the slender robot can move in different ways, it includes single and dual-degree-of-freedom joints like deflector, rotary and sliding and all using a tendon-sheath for flexibility.

Deflection joint module: The deflection joint, as seen in Fig. 6.4a, has one movement direction. It links the moving part to the base using a rotating pulley. Instead of a traditional bearing, a copper sleeve is used to facilitate its slow movement, making the structure more compact and straightforward. Two deflector joints can be installed in parallel (Fig. 6.4b) or rotated by 90 °C for vertical installation (Fig. 6.4c). The deflector joint has a diameter of 40 mm, a length of 28 mm, and can move up to 63 °C. The joints are driven using a double tendon-sheath transmission system.

Rotary joint module: The rotary joint allows the joint to pivot along the centerline of the robot and has one movement direction. It primarily includes bearings, a bearing cover, connecting plate, central shaft, pulley shaft, and pulley shaft bracket. As depicted in Fig. 6.5, the joint is driven by a double tendon-sheath mechanism. specifically, this joint uses two fixed pulleys to alter the rope direction. The length of the joint is 63 mm, the diameter is 40 mm, and the rotation range is 90 °C.

Sliding joint module: The sliding joint moves the end-effector along the axis of the robot and has four parts: the base, movable section, spring, and guide key (Fig. 6.6). Two tendon-sheath mechanism are arranged symmetrically on both sides

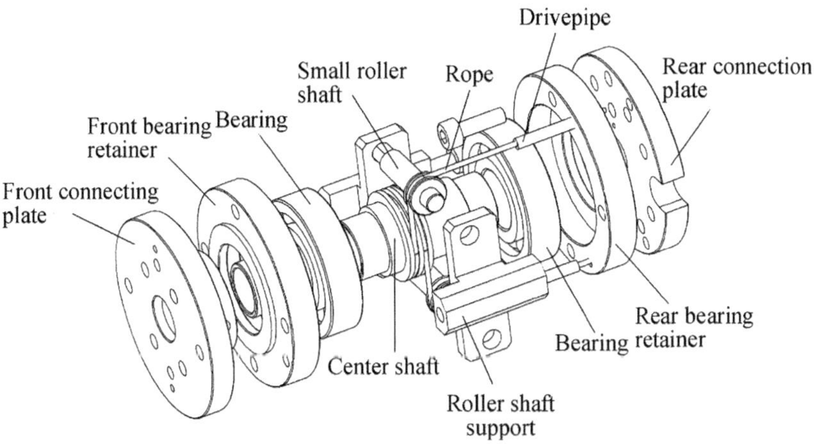

Fig. 6.5 Schematic diagram of the structure of the rotary joint module [17]

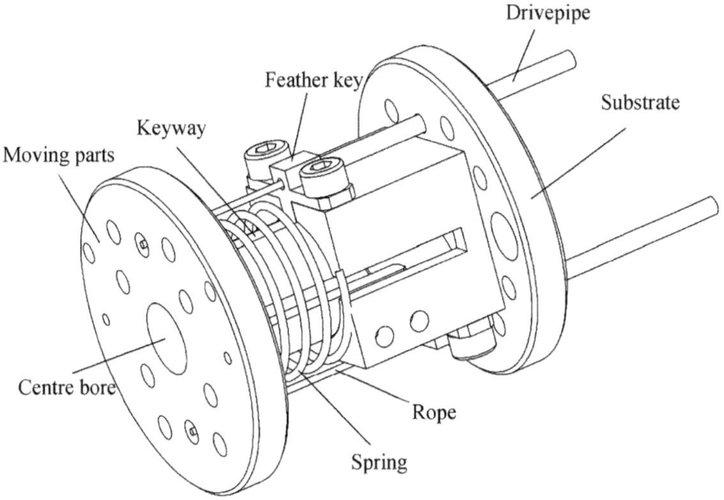

Fig. 6.6 Schematic diagram of the structure of the sliding joint module [17]

of the joint and are driven by two servo motors respectively. When the two servos move simultaneously at the same speed, the movable member slides relative to the base and the spring on the outside of the tab is compressed. When the servos relax the ropes, the spring return force can drive the movable member to reset. The diameter of the joint is 40 mm, the length is 54 mm, and the range of joint movement is 10 mm.

End-effector module: The end-effector, shown in Fig. 6.7, is a compact two-finger gripper with a double parallelogram design that provides effective gripping. The rope connects to the gripper's central shaft. Pulling the rope causes the fingers

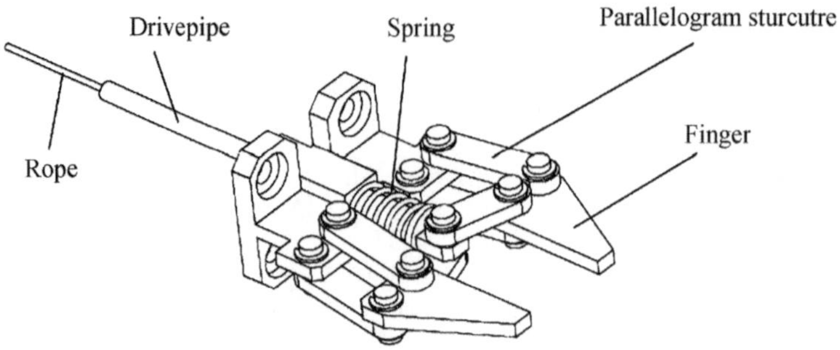

Fig. 6.7 Schematic diagram of end-effector module structure [17]

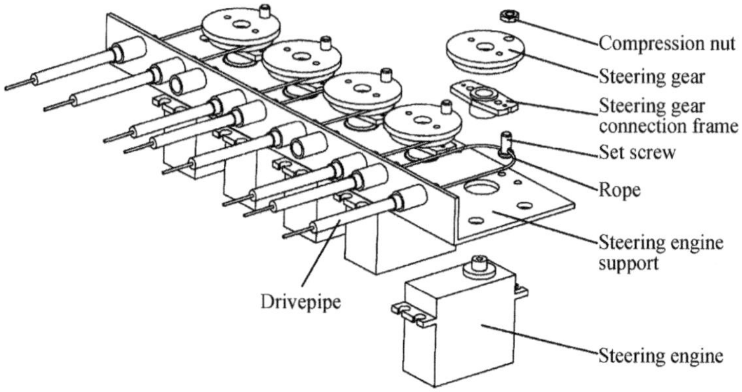

Fig. 6.8 Drive unit structure [17]

to close by compressing a built-in spring on the shaft. Releasing the rope allows the spring to open the fingers. The gripper measures a maximum of 34 mm × 44 mm × 20 mm and can open and close within a range of 12 mm.

Joint drive unit: Since the designed joint movements are within a limited distance and angle range, the rudder is selected as the drive element for each joint from the aspect of economy and feasibility. Each joint is driven by the rudder, and the driving end is shown in Fig. 6.8. Because of the use of the tendon-sheath mechanism, the driving end can drive the end effector at a long distance. The prototypes of these joints are shown in Fig. 6.9.

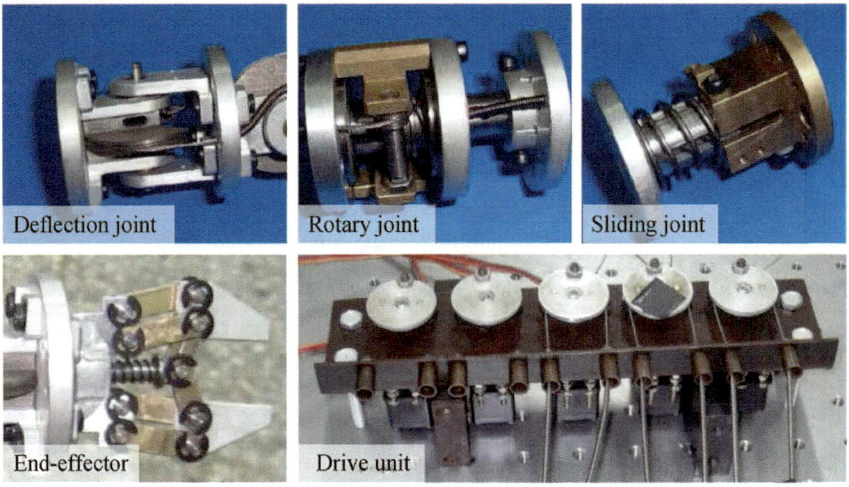

Fig. 6.9 Prototypes of slender link series [17]

6.4 Kinematics Analysis of Micro Tip Manipulators

6.4.1 Modular Joint Kinematics Analysis Based on Screw Theory

(1) Basis of screw theory

Forward kinematics determine the end shape of the robot based on its connected parts. Since the head joints of robots are designed in a modular way and have a variety of configurations, we need to make the relative kinematic analysis. Early in the 19th century, Chasles and Poinsot had related spinor theory in their works, and later Robert S. Ball expanded on their ideas, providing a comprehensive spinor theory in 1900. Richard M et al. later updated this theory using linear algebra and matrix groups. Their approach focuses on representing rigid body motion using specific coordinates and transforming kinematic spirals into relevant motions using matrix transformations. By using this representation, the kinematics of a tandem robot can be simplified into a straightforward exponential form.

Chasles proved that the motion of a rigid body from one position to another can be obtained by a rotation around a certain line plus a movement along that line, and this combined motion is called spiral motion, and the infinitesimal amount of spiral motion is the motion spiral. The end shape space of the rigid body motion transformation is$SE(3)$, which is a special Euclidean group and is the product space of R^3 and $SO(3)$, denoted as$SE(3) = \left\{(\boldsymbol{p}, \boldsymbol{R}) : \boldsymbol{p} \in \mathfrak{R}^3, \boldsymbol{R} \in SO(3)\right\} = \mathfrak{R}^3 \times SO(3)$.

If a rigid body moves from coordinate system A to coordinate system B, its positional transformation $\boldsymbol{g_{ab}} \in SE(3)$, can be expressed in terms of the chi-square matrix as

$$g_{ab} = \begin{bmatrix} \boldsymbol{R}_{ab} & \boldsymbol{p}_{ab} \\ 0 & 1 \end{bmatrix}_{4\times4} \tag{6.4.1}$$

where $\boldsymbol{R}_{ab} \in SO(3)$ represents the rotational transformation of a rigid body and $\boldsymbol{p}_{ab} \in \mathfrak{R}^3$ represents the translational transformation of a rigid body.

$SE(3)$ the Lie algebra of $se(3)$, the elements of which are said to be the kinematic spins $\hat{\boldsymbol{\xi}}$, is expressed in chi-square coordinates as

$$\hat{\boldsymbol{\xi}} = \begin{bmatrix} \hat{\boldsymbol{\omega}} & \boldsymbol{v} \\ 0 & 0 \end{bmatrix} \in \mathfrak{R}^{4\times4} \tag{6.4.2}$$

where $\omega = \left(\omega_x, \omega_y, \omega_z\right)$ is the unit direction vector of the axis of rotation, $\|\boldsymbol{\omega}\| = 1$, and

$$\hat{\boldsymbol{\omega}} = \begin{bmatrix} 0 & -\omega_z & \omega_y \\ \omega_z & 0 & -\omega_x \\ -\omega_y & \omega_x & 0 \end{bmatrix} \in so(3) \tag{6.4.3}$$

$\boldsymbol{v} \in \boldsymbol{R}^3$ is the unit vector in the direction of movement. In order to represent the motion spin in 6-dimensional vectors, the V operator is introduced

$$\begin{bmatrix} \hat{\boldsymbol{\omega}} & \boldsymbol{v} \\ 0 & 0 \end{bmatrix}^V = \begin{bmatrix} \boldsymbol{v} \\ \boldsymbol{\omega} \end{bmatrix} \tag{6.4.4}$$

and call the rotational coordinates of $\xi \triangleq (v, \omega)$ the motion of ξ. For a rotating joint $\boldsymbol{\xi} = (-\boldsymbol{\omega} \times \boldsymbol{q}, \boldsymbol{\omega})$, where $\boldsymbol{q}$ is any point on the axis of rotation; for a moving joint $\boldsymbol{\xi} = (\boldsymbol{v}, 0)$.

For a given $\hat{\boldsymbol{\xi}} \in se(3)$ and $\theta \in \boldsymbol{R}$, $e^{\hat{\xi}\theta}$ the elements of indexed to $SE(3)$ can be used to describe the transformation of the rigid body from the start to the final bit shape,

$$g_{ab}(\theta) = e^{\hat{\xi}\theta} g_{ab}(0) \tag{6.4.5}$$

where $\hat{\boldsymbol{\xi}}_\theta$ is calculated as follows:

$$e^{\hat{\xi}\theta} = \begin{cases} \begin{bmatrix} \boldsymbol{I} & \boldsymbol{v}\theta \\ 0 & 1 \end{bmatrix} & , \boldsymbol{\omega} = 0 \\ \begin{bmatrix} e^{\hat{\omega}\theta} & \left(\boldsymbol{I} - e^{\hat{\omega}\theta}\right)(\boldsymbol{\omega} \times \boldsymbol{v}) + \boldsymbol{\omega}\boldsymbol{\omega}^T \boldsymbol{v}\theta \\ 0 & 1 \end{bmatrix} & , \boldsymbol{\omega} \neq 0 \end{cases} \tag{6.4.6}$$

When there are multiple joints in series, combining the motions of the joints with each other gives the total robot kinematics with the positive solution exponential product equation.

$$\boldsymbol{g}_{ab}(\boldsymbol{\theta}) = e^{\hat{\xi}_1\theta_1}e^{\hat{\xi}_2\theta_2}\cdots e^{\hat{\xi}_n\theta_n}\boldsymbol{g}_{ab}(0) \tag{6.4.7}$$

where $\hat{\boldsymbol{\xi}}_i$, $i = 1, 2, \ldots, n$ is the i joint motion spin volume.

$e^{\hat{\xi}_i\theta_i}$, $i = 1, 2, \ldots, n$ is i joint spin volume movement.

$\boldsymbol{g}_{ab}(0)$ is the relative position of the coordinate system B with respect to the base coordinate system A at the time of the reference position shape.

$\boldsymbol{g}_{ab}(\boldsymbol{\theta})$ is the relative position of coordinate system B with respect to the base coordinate system A, given the joint variable $\boldsymbol{\theta} = (\theta_1, \theta_2, \ldots, \theta_n))$.

$\boldsymbol{\xi}_i$ must be numbered sequentially starting from the base, but the rotation and movement of the coordinate system given by $\boldsymbol{g}_{ab}(\boldsymbol{\theta})$ are independent of the actual order. From $F_{inv}(t) = \begin{cases} \frac{F_d(t)}{Q_p}, & \dot{F}_d(t) > 0 \\ \frac{F_d(t)}{Q_r}, & \dot{F}_d(t) < 0 \\ F_{inv}\left(t^-\right), & \dot{F}_d(t) = 0 \end{cases}$. Using the exponential product offers a clearer and more straightforward way to describe robot joint movements than the D-H parametric method. When defining the kinematics of the robot, we only need to focus on the axis direction of the joint rotation, not its specific configuration. This method uses kinematic rotation coordinates to give an overall picture of the rigid body's movement.

(2) Modular joint kinematics analysis

Consider a 5-degree-of-freedom rigid joint of the head as shown follow, which consists of a rotary joint (joint 1), a deflector joint (joint 2), a sliding joint (joint 3) and two-degree-of-freedom joints (joint 4 and joint 5) in the configuration of RRTRR. Here, the rotary joint is depicted as a cylinder, and the moving joint appears as a rectangular shape. The base coordinate system S and the end tool coordinate system T shown in Fig. 6.10 are established by taking $\boldsymbol{\theta} = 0$ corresponding to the fully extended shape shown in the figure.

When $\boldsymbol{\theta} = 0$ the transformation of the base and tool coordinate systems in time is

$$g_{st}(0) = \begin{bmatrix} \boldsymbol{I} & \begin{bmatrix} 0 \\ l_1 + l_2 + l_3 + l_4 \\ 0 \end{bmatrix} \\ 0 & 1 \end{bmatrix} \tag{6.4.8}$$

The unit vectors for each axis of rotation are

$$\boldsymbol{\omega}_1 = \boldsymbol{\omega}_5 = \begin{bmatrix} 0 \\ 1 \\ 0 \end{bmatrix}, \quad \boldsymbol{\omega}_2 = \begin{bmatrix} 0 \\ 0 \\ 1 \end{bmatrix}, \quad \boldsymbol{\omega}_4 = \begin{bmatrix} 1 \\ 0 \\ 0 \end{bmatrix} \tag{6.4.9}$$

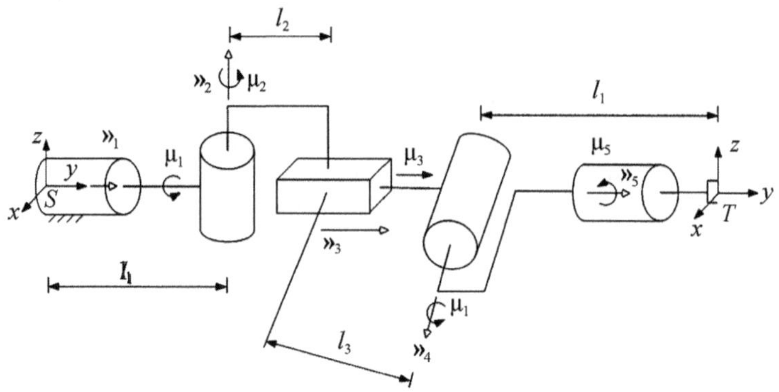

Fig. 6.10 5-degree-of-freedom head joint consisting of rotary, deflection, sliding joint module and two-degree-of-freedom joints

The points on the rotation axis are taken as

$$\boldsymbol{q}_1=\begin{bmatrix}0\\0\\0\end{bmatrix},\quad \boldsymbol{q}_2=\begin{bmatrix}0\\l_1\\0\end{bmatrix},\quad \boldsymbol{q}_4=\begin{bmatrix}0\\l_1+l_2+l_3\\0\end{bmatrix},\quad \boldsymbol{q}_5=\begin{bmatrix}0\\l_1+l_2+l_3+l_4\\0\end{bmatrix} \tag{6.4.10}$$

The corresponding kinematic spins can be obtained from $\boldsymbol{\xi}_i=\begin{bmatrix}-\boldsymbol{\omega}_i\times\boldsymbol{q}_i & \boldsymbol{\omega}_i\end{bmatrix}^T$ as

$$\boldsymbol{\xi}_1=\begin{bmatrix}0\\0\\0\\0\\1\\0\end{bmatrix},\quad \boldsymbol{\xi}_2=\begin{bmatrix}l_1\\0\\0\\0\\0\\1\end{bmatrix},\quad \boldsymbol{\xi}_4=\begin{bmatrix}0\\0\\-(l_1+l_2+l_3)\\1\\0\\0\end{bmatrix},\quad \boldsymbol{\xi}_5=\begin{bmatrix}0\\0\\0\\0\\1\\0\end{bmatrix} \tag{6.4.11}$$

Taking the mobile joint along the y axis direction, the corresponding rotational amount of motion is:

$$\boldsymbol{\xi}_3=\begin{bmatrix}\boldsymbol{v}_3\\0\end{bmatrix}=\begin{bmatrix}0\\1\\0\\0\\0\\0\end{bmatrix} \tag{6.4.12}$$

Using Rodrigues' formula and $F_{\text{out}}(t) = \mathcal{BL}(F_{\text{in}}(t)) = \begin{cases} F_{\text{in}}(t)Q_p, \text{ if } \lambda < 0 \\ F_{\text{in}}(t)Q_r, \text{ if } \lambda > 0 \\ F_{\text{out}}(t^-), \text{ otherwise} \end{cases}$,

the exponent of each kinematic spin can be calculated as

$$e^{\hat{\xi}_1\theta_1} = \begin{bmatrix} \cos\theta_1 & 0 & \sin\theta_1 & 0 \\ 0 & 1 & 0 & 0 \\ -\sin\theta_1 & 0 & \cos\theta_1 & 0 \\ 0 & 0 & 0 & 1 \end{bmatrix} \tag{6.4.13}$$

$$e^{\hat{\xi}_2\theta_2} = \begin{bmatrix} \cos\theta_2 & -\sin\theta_2 & 0 & l_1\sin\theta_2 \\ \sin\theta_2 & \cos\theta_2 & 0 & l_1(1-\cos\theta_2) \\ 0 & 0 & 1 & 0 \\ 0 & 0 & 0 & 1 \end{bmatrix} \tag{6.4.14}$$

$$e^{\hat{\xi}_3\theta_3} = \begin{bmatrix} 1 & 0 & 0 & \\ 0 & 1 & 0 & \theta_3 \\ 0 & 0 & 0 & \\ 0 & 0 & 1 & \end{bmatrix} \tag{6.4.15}$$

$$e^{\hat{\xi}_4\theta_4} = \begin{bmatrix} 1 & 0 & 0 & 0 \\ 0 & \cos\theta_4 & -\sin\theta_4 & (l_1+l_2+l_3)(1-\cos\theta_4) \\ 0 & \sin\theta_4 & \cos\theta_4 & -(l_1+l_2+l_3)\sin\theta_4 \\ 0 & 0 & 0 & 1 \end{bmatrix} \tag{6.4.16}$$

$$e^{\hat{\xi}_5\theta_5} = \begin{bmatrix} \cos\theta_5 & 0 & \sin\theta_5 & 0 \\ 0 & 1 & 0 & 0 \\ -\sin\theta_5 & 0 & \cos\theta_5 & 0 \\ 0 & 0 & 0 & 1 \end{bmatrix} \tag{6.4.17}$$

The flush transformation matrix of this configuration is

$$g_{st}(\boldsymbol{\theta}) = \begin{bmatrix} \boldsymbol{R}(\boldsymbol{\theta}) & \boldsymbol{p}(\boldsymbol{\theta}) \\ 0 & 1 \end{bmatrix} \tag{6.4.18}$$

Among them

$$\boldsymbol{R}(\boldsymbol{\theta}) = \begin{bmatrix} r_{11} & r_{12} & r_{13} \\ r_{21} & r_{22} & r_{23} \\ r_{31} & r_{32} & r_{33} \end{bmatrix}, \quad \boldsymbol{p}(\boldsymbol{\theta}) = \begin{bmatrix} p_{11} \\ p_{21} \\ p_{31} \end{bmatrix}$$

If we use the notation: $c_i = \cos\theta_i, \quad s_i = \sin\theta_i$, then the parameters of the transformation matrix correspond as follows.

$$
\begin{aligned}
r_{11} &= c_1c_2c_5 - s_5\left(c_4s_1 + c_1s_2s_4\right)\\
r_{12} &= s_1s_4 - c_1c_4s_2\\
r_{13} &= c_5\left(c_4s_1 + c_1s_2s_4\right) + c_1c_2s_5\\
r_{21} &= c_5s_2 + c_2s_4s_5\\
r_{22} &= c_2c_4\\
r_{23} &= s_2s_5 - c_2c_5s_4\\
r_{31} &= -s_5\left(c_1c_4 - s_1s_2s_4\right) - c_2c_5s_1\\
r_{32} &= c_1s_4 + c_4s_1s_2\\
r_{33} &= c_5\left(c_1c_4 - s_1s_2s_4\right) - c_2s_1s_5\\
p_{11} &= \left(l_1 + l_2 + l_3 + l_4\right)\left(s_1s_4 - c_1c_4s_2\right) + \left(l_1 - \theta_3\right)c_1s_2\\
&\quad + \left(l_1 + l_2 + l_3\right)\left[c_1s_2\left(c_4 - 1\right) - s_1s_4\right]\\
p_{12} &= \theta_3c_2 - l_1\left(c_2 - 1\right) - c_2\left(c_4 - 1\right)\left(l_1 + l_2 + l_3\right) + c_2c_4\left(l_1 + l_2 + l_3 + l_4\right)\\
p_{13} &= s_1s_2\left(\theta_3 - l_1\right) - \left(l_1 + l_2 + l_3\right)\left[c_1s_4 + s_1s_2\left(c_4 - 1\right)\right]\\
&\quad + \left(l_1 + l_2 + l_3 + l_4\right)\left(c_1s_4 + c_4s_1s_2\right)
\end{aligned}
$$

6.4.2 Workspace Analysis of the Slender Link Series

(1) Principle of Monte Carlo analysis workspace

The workspace indicates the working area that the robot end-effector can access. Here the robot consists of a head joint and a flexible body, and the flexible part cannot be controlled directly, so here the robot workspace refers to the workspace of the rigid joint consisting of the head joint, i.e., the range in which the end-effector can operate after the robot reaches the workplace. Modular reconfigurability is characterized by flexible customization of degrees of freedom and workspace, i.e., the type of joints and the connection method can be changed flexibly. In order to facilitate the solution and reduce the tedious manual computation, the method of visualizing and determining the boundaries of the robot workspace under arbitrary configurations is studied here.

Once the structural parameters of the modular joints of the head have been obtained from the analysis in the front subsection, their workspace can be represented by the generalized coordinates θ (joint variables). However, due to the limitations of the actual joint structure and actuators, all joints can only move within a specific range, $\theta_{\min} \leqslant \boldsymbol{\theta} \leqslant \boldsymbol{\theta}_{\max}$, where $\theta_{\max}$ and $\theta_{\min}$ is the upper and lower limit of θ respectively. The coordinate origin of the tool coordinate system is selected as O_T, O_T and the position of the end P_T coincide, and the set of points it can reach is the workspace of the head joint, and the workspace is represented by the set:

$$W(P) = \{\boldsymbol{P}_T \mid \boldsymbol{\theta}_{\min} \leqslant \boldsymbol{\theta} \leqslant \boldsymbol{\theta}_{\max}\} \tag{6.4.19}$$

Table 6.1 Parameters of spin measure coordinates of modular head joints with 5 degrees of freedom

Joint i	Rotational coordinates ξ_i	Range of nodal variables θ_i
1	$\begin{bmatrix} 0\,0\,0\,0\,1\,0 \end{bmatrix}^T$	$[-90°, +90°]$
2	$\begin{bmatrix} 0.077\ 0\,0\,0\,0\,1 \end{bmatrix}^T$	$[-63°, +63°]$
3	$\begin{bmatrix} 0\,1\,0\,0\,0\,0 \end{bmatrix}^T$	$[-0.01, 0]$
4	$\begin{bmatrix} 0\,0\ 0.178\ 1\,0\,0 \end{bmatrix}^T$	$[-90°, +90°]$
5	$\begin{bmatrix} 0\,0\,0\,0\,1\,0 \end{bmatrix}^T$ $[-90°, +90°]$	

Currently, there are mainly analytical, graphical and numerical methods for solving workspaces. The analytical method is complex and unintuitive, and is generally limited to the analysis of the workspace of robots with less than three degrees of freedom. The graphical method is simple and intuitive, but it is also limited by the number of joints and requires grouping when the degrees of freedom are large. The numerical method is generally too computationally intensive, especially when the boundary surface is concave, and reliability is not always guaranteed.

One approach to numerical analysis is the Monte Carlo method, which utilizes random sampling to solve mathematical challenges. By randomly exploring generalized coordinate values within a set range, this method can define the robot workspace. By employing pseudo-random numbers like $(\cdot)$ between 0 and 1, each joint variable's value can be systematically determined.

$$\theta_i^k = \theta_{i\,\min} + (\theta_{i\,\max} - \theta_{i\,\min})\,\mathrm{RAND} \quad (i = 1, 2, \ldots, n;\, k = 1, 2, \ldots, \lambda) \tag{6.4.20}$$

where i is the joint number and λ is the total number of samples of the generated random quantities (Table 6.1).

Using the Monte Carlo method and considering the motion range of each joint from 6.1, we sampled 99,999,999 points. This process took 7.13 s to visualize the workspace of the head joints, depicted as a point cloud in Fig. 6.11. However, while this method is straightforward, the resulting point cloud doesn't clearly define the workspace boundaries, especially its inner limits. Because this representation of the workspace of the head joint by Monte Carlo point cloud map alone is neither intuitive nor complete. In the following sections, we'll explore methods to clarify both the inner and outer boundaries of this workspace using the Monte Carlo point cloud.

(2) Finding the workspace boundary

The Monte Carlo point cloud map represents a three-dimensional workspace. To simplify and identify its inner and outer boundaries, we transition from a 3D boundary search to multiple 2D planes using parallel sections. Within these 2D planes, we

Fig. 6.11 Workspace Monte Carlo point cloud [17]

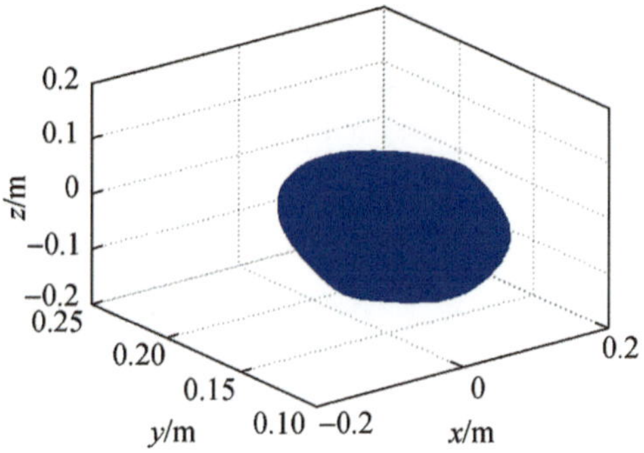

continue the boundary search using parallel lines. As shown in Fig. 6.12, the boundary line in these sections can have both elevated and concave parts. The method of finding both is not the same, and the specific steps of boundary point finding are as follows.

1. Sorting the coordinates of the Monte Carlo cloud z to obtain the limit points $z_{\min}$ and $z_{\max}$. In the direction of the workspace z, two planes $z_u = z_{\max}$ and $z_d = z_{\max} - \Delta z$ are used to intercept the workspace starting from $z_{\max}$, where Δz is the step size of the interception, which can be adjusted according to the accuracy requirements. Since it is relatively small, it can be assumed that the workspace point clouds in both planes lie on the same plane.
2. The Monte Carlo point cloud in this interface is sorted by y coordinates to obtain the limit points $y_{\min}$ and $y_{\max}$ in the direction of y in this section. Then two lines $y_u = y_{\max}$ and $y_d = y_{\max} - \Delta y$ are used to intercept the workspace starting from $z_{\max}$, where Δy is the intercept step size, which can be adjusted according to the accuracy requirement. Since it is relatively small, it can be assumed that the workspace point clouds in both planes lie on the same line.
3. The point clouds intercepted by the lines $y_u = y_{\max}$ and $y_d = y_{\max} - \Delta y$ are sorted by x coordinates, and the resulting $x_{\min}$ and $x_{\max}$ are the x coordinates of the convex boundary points.
4. The sorted x coordinates are subtracted in order, i.e. $x_2 - x_1, x_3 - x_2, \ldots, x_n - x_{n-1}$, where $x_i > x_{i-1}$. If where $x_j - x_{j-1} > \sigma$, σ is a relatively large number, then consider x_j, x_{j-1} as the x coordinates of the two concave boundary points. The reason is that the ordered x coordinates will mutate only if there is a concave boundary line, and the place where they mutate is exactly the concave boundary point.

Within a Δy step, there is one and only one pair of convex boundary points. However, concave boundary points might be present or absent, and if they are, there could be multiple pairs. Once we determine the boundary lines for one section, we repeat the process in Δy increments until we map out the entire workspace boundary. Using this method, the Monte Carlo point cloud in Fig. 6.11 is used to find the boundary lines, and the shapes and projections of the convex and concave boundaries are shown in Figs. 6.13 and 6.14.

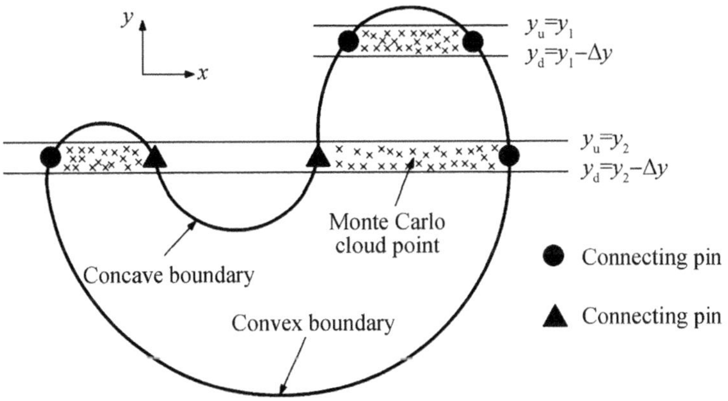

Fig. 6.12 Workspace Monte Carlo point cloud [17]

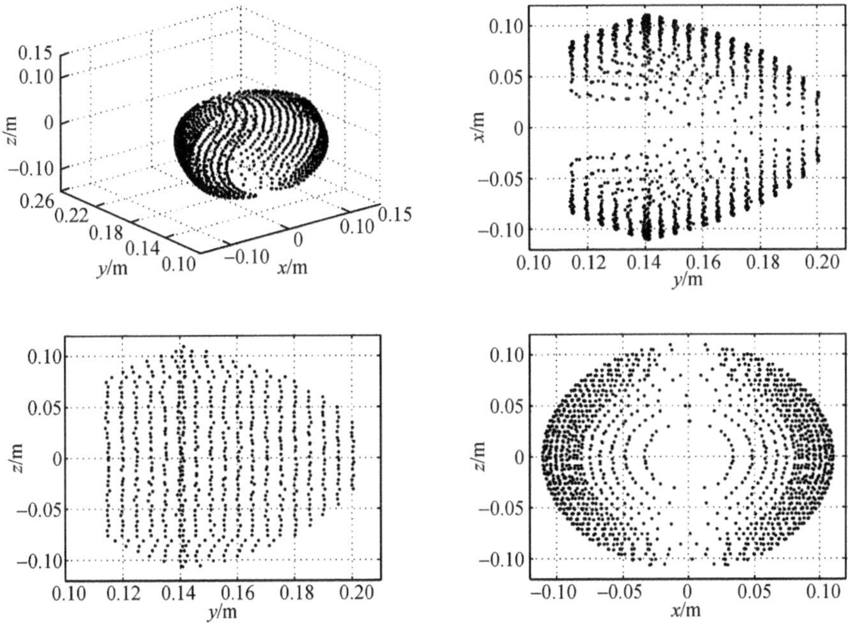

Fig. 6.13 Projection plot of workspace convex boundary [17]

6.5 Compliant Design of Micro Tip Operators

In this chapter, we discuss a flexible head actuator for robots, including a clamping joint, a two-axis bending joint, and a rotation joint [18]. Based on global research and practical needs, we've defined how these joints move, detailed in Table 6.2.

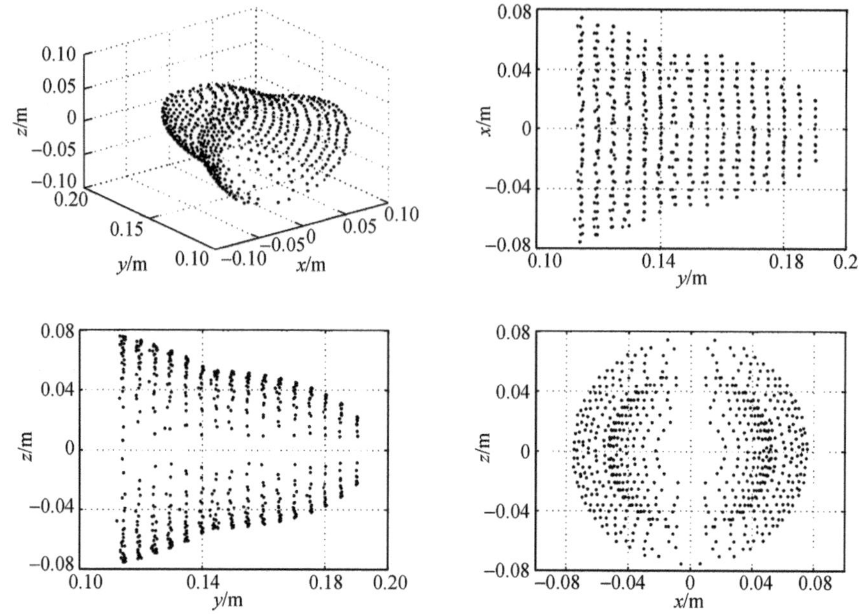

Fig. 6.14 Projection plot of workspace concave boundary [17]

Table 6.2 Head actuator range of motion

Serial number	Joint name	Range of motion
(1)	Clamping joints	Clamping of objects less than 20 mm in diameter
(2)	Bending joints	Capable of bending 45° in any direction
(3)	Slewing joints	Capable of forward/reverse rotation of 45°

For effective search, rescue, and detection tasks, the robot must interact with its target. At the front end of the robot, the clamping joint handles gripping and releasing objects. Since our bending joint moves in one direction, we use two in sequence to allow bending in any direction. Behind this bending joint is a rotation joint around the main axis. the rotation joint has five parts in a series to ensure the robot can move as needed for its tasks.

The elastic manipulator is presented in Fig. 6.15, it has three degrees of freedom (DOFs) for adjusting orientation and posture, and an additional DOF for grasping. As can be observed, the tendon-sheath driven manipulator can be basically divided into three parts: the gripper, the rotary component and the bending component. When the tendon connected to the inner beam of the gripper is pulled, the gripper closes to

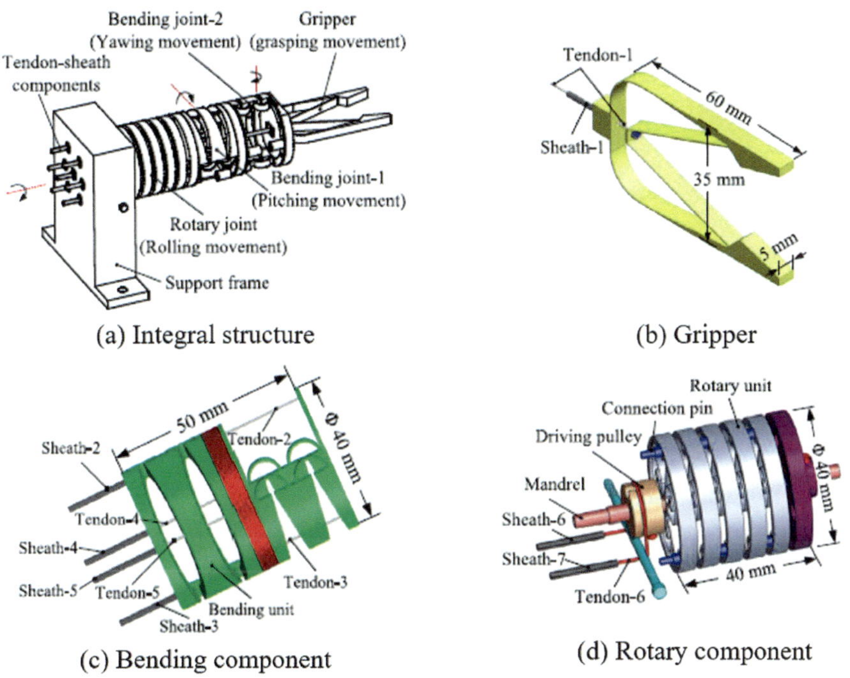

Fig. 6.15 CAD model of the compliant manipulator (Reprinted from [19] with permission from Elsevier)

grasp an object. Conversely, due to its elastic nature, the gripper automatically opens when the tendon is released.

As shown in Fig. 6.15b, the gripper has two symmetric gripper arms. The sheath is fixed at the rectangular boss at the bottom of the manipulator, and the tendon is connected with the middle vertex position of V beam. The distance between the two gripper arms will decrease when the tendon is pulled. As a result, the manipulator will clamp the target object.

The rotary joint plays a significant role during rescue operation. When the tendon fixed on the groove of driving pulley is pulled, the rotary component is able to perform the rolling movement in the range of ±90°. The bending component consists of two C-shape compliant bending units, which are mutually orthogonal and can perform the yawing movement and the pitching movement in the range of ±45° respectively. Figure 6.16 illustrates the photographs of the compliant manipulator performing different actions.

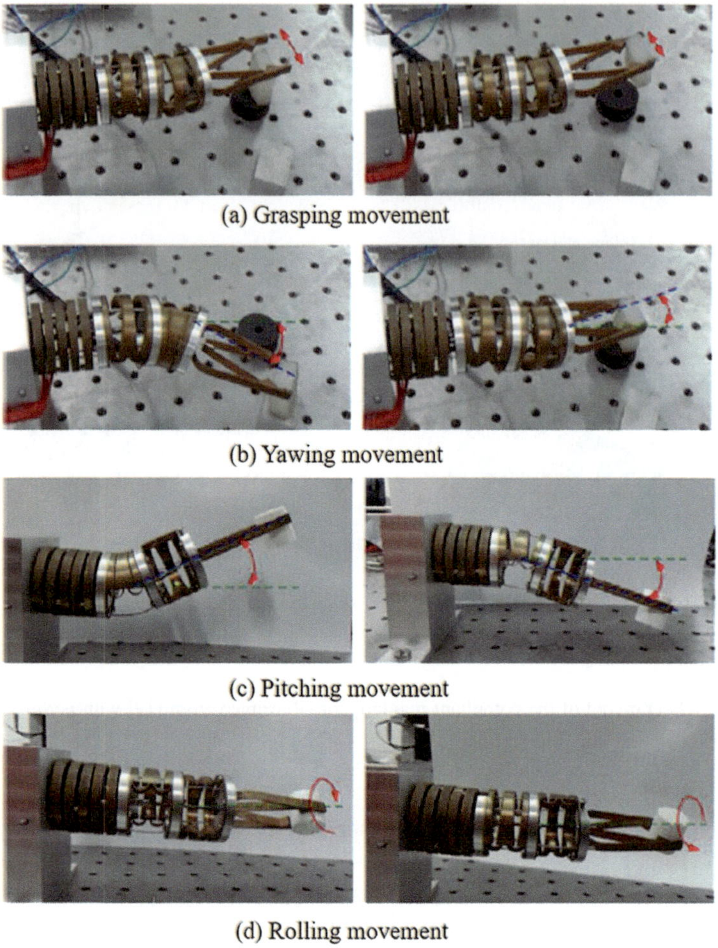

Fig. 6.16 Photographs of the compliant manipulator performing different actions (Reprinted from [19] with permission from Elsevier)

References

1. Wang M, Palmer D, Dong X, et al. Design and development of a slender dual-structure continuum robot for in-situ aero engine repair. In: 2018 IEEE/RSJ international conference on intelligent robots and systems (IROS). IEEE, pp 5648–5653
2. Troncoso D, Robles-Linares J, Russo M et al (2022) A continuum robot for remote applications: from industrial to medical surgery with slender continuum robots. IEEE Robot Autom Mag
3. Racioppo P, Ben-Tzvi P (2019) Design and control of a cable-driven articulated modular snake robot. IEEE/ASME Trans Mechatron 24(3):893–901
4. Hannan M, Walker I (2003) Kinematics and the implementation of an elephant's trunk manipulator and other continuum style robots. J Robot Syst 20(2):45–63

5. Sharon A, Hogan N, Hardt D (1993) The macro/micro manipulator: an improved architecture for robot control. Robot Comput Integr Manuf 10(3):209–222
6. Uzunoglu E, Tatlicioglu E, Dede M (2021) A multi-priority controller for industrial macro-micro manipulation. Robotica 39(2):217–232
7. Khatib O (1988) Reduced effective inertia in macro-/mini-manipulator systems. In: Proceedings of ACC, pp 2140–2147
8. Arifin A, Ang M, Lai C et al (2013) General framework of the force and compliant motion control for macro mini manipulator. In: 2013 IEEE/ASME international conference on advanced intelligent mechatronics. IEEE, pp 949–954
9. Yoshikawa T, Harada K, Matsumoto A (1996) Hybrid position/force control of flexible-macro/rigid-micro manipulator systems. IEEE Trans Robot Autom 12(4):633–640
10. Chen H, Li J, Xing G et al (2010) Trajectory tracking control of a macro-micro welding robot based on the vision navigation. In: 2010 IEEE international conference on robotics and biomimetics. IEEE, pp 667–672
11. Ma Z, Hong G, Ang M et al (2016) Design and control of an end-effector module for industrial finishing applications. In: 2016 IEEE international conference on advanced intelligent mechatronics (AIM). IEEE, pp 339–344
12. Uzunoglu E, Dede M, Kiper G (2016) Trajectory planning for a planar macro-micro manipulator of a laser-cutting machine. Ind Robot Int J 43(5):513–523
13. Dede MIC, Kiper G, Uzunoglu E (2016) A macro-micro mechanism design for laser cutting process
14. Liu J, Tong Y, Liu J (2021) Review of snake robots in constrained environments. Robot Auton Syst 141:103785
15. Russo M, Sadati S, Dong X et al (2023) Continuum robots: an overview. Adv Intell Syst 5(5):2200367
16. Yim M, Zhang Y, Duff D (2002) Modular robots. IEEE Spectr 39(2):30–34
17. Chen L (2013) Modeling and control of tendon-sheath transmission system and its application in long and slender robot for searching and inspection. Southeast University
18. Wang K (2014) Design and preliminary experiments on soft manipulation of tendon-sheath driven compliant robot. Southeast University
19. Wu Q, Wang X, Chen B et al (2017) Development and hybrid force/position control of a compliant rescue manipulator. Mechatronics 46:143–153. https://doi.org/10.1016/j.mechatronics.2017.08.003

Chapter 7
Geometry Estimation of Tendon-Sheath Transmitted Slender Robots

Abstract Slender robots are designed to work in narrow spaces, such as the disaster, the industrial equipment and the minimally invasive surgery. However, when these robots navigate such confined areas, it's challenging to figure out the space shape and the robot position inside it. As a result, we cannot know the spatial position of the end effector inside the space, and thus cannot achieve localization of the robot. In order to solve this problem, it is necessary to consider the shape reconstruction and localization of the slender robot in confined environment. This chapter introduces four common and innovative reconstruction methods for different structure types and application scenarios of slender robots, including the traditional D-H method for slender robot with multiple linked segments, the near-end sensing and geometry estimation method for robots without end sensor feedback, the distal-end sensing method and the fiber brag grating sensing method for estimating the geometry shape. The body shape reconstruction is crucial for the control and application of slender robots, this chapter covers both fundamental and advanced techniques in this area.

7.1 Body Shape Reconstruction from Joint Angles of Series Link

The ability of slender robots to negotiate curved paths is a particularly critical aspect of their operational capabilities. The positioning and body shape reconstruction of such robots is a key issue in controlling them effectively [1, 2]. Moreover, the transmission characteristics of the tendon-sheath system are strongly influenced by the tendon bending configuration. This configuration, in turn, is heavily influenced by the configuration of the robot joints. As a result, accurate reconstruction of these joint configurations is important for precise control of slender robots.

To address these challenges, this section introduces a 3D reconstruction method for an elongated robot with a multi-joint manipulator. This method can calculate the configuration of the sheath based on the rotation angles of the robot joints. By accurately reconstructing the joint configurations, this method aims to improve the control accuracy of slender robots and enhance their operational capabilities.

X. Wang and Q. Zhang, *Tendon-Sheath Transmitted Light and Slender Robots*,
https://doi.org/10.1007/978-981-96-5581-6_7

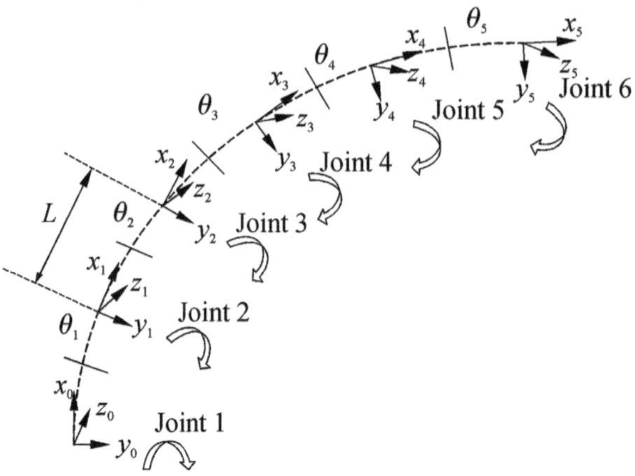

Fig. 7.1 Coordinate system of the steering head

As shown in Fig. 7.1, the D-H (Denavit-Hartenberg) method was utilized to establish the spatial coordinate system for steering head. The length of the link rods between these joints are L_1, L_2, L_3, L_4, L_5 and L_6, which can be measured in the steering head; the rotation angles of the six joints are $\theta_1, \theta_2, \theta_3, \theta_4, \theta_5$, and θ_6; A, B, C, D, and E are five reference points. The D-H parameter table can be obtained as follows.

$$ {}^{i-1}_{\ \ i}T = \begin{bmatrix} c\theta_i & -s\theta_i & 0 & a_{i-1} \\ s\theta_i c\alpha_{i-1} & c\theta_i c\alpha_{i-1} & -s\alpha_{i-1} & -s\alpha_{i-1} d_i \\ s\theta_i s\alpha_{i-1} & c\theta_i s\alpha_{i-1} & c\alpha_{i-1} & c\alpha_{i-1} d_i \\ 0 & 0 & 0 & 1 \end{bmatrix} \tag{7.1.1} $$

Substitute the D-H parameters from Table 7.1 into Eq. 7.1.1, the transformation matrix between each coordinate system can be calculated as follows

Table 7.1 D-H parameters of the slender robot with multiple links

Joint index	a_{i-1} (cm)	α_{i-1}	d_i (cm)	θ_i
1	0.00	0	0	θ_1
2	6.00	$-\pi/2$	0	θ_2
3	6.00	$\pi/2$	0	θ_3
4	6.00	$-\pi/2$	13.0	θ_4
5	6.00	$\pi/2$	0	θ_5
6	6.00	$-\pi/2$	10.0	θ_6

$$
{}^{0}_{1}T = \begin{bmatrix} c\theta_1 & -s\theta_1 & 0 & 0 \\ s\theta_1 & c\theta_1 & 0 & 0 \\ 0 & 0 & 1 & 0 \\ 0 & 0 & 0 & 1 \end{bmatrix} \tag{7.1.2}
$$

$$
{}^{l}_{2}T = \begin{bmatrix} c\theta_2 & -s\theta_2 & 0 & 6 \\ 0 & 0 & -1 & 0 \\ s\theta_2 & c\theta_2 & 0 & 0 \\ 0 & 0 & 0 & 1 \end{bmatrix} \tag{7.1.3}
$$

$$
{}^{2}_{3}T = \begin{bmatrix} c\theta_3 & -s\theta_3 & 0 & 6 \\ 0 & 0 & 1 & 0 \\ -s\theta_3 & -c\theta_3 & 0 & 0 \\ 0 & 0 & 0 & 1 \end{bmatrix} \tag{7.1.4}
$$

$$
{}^{3}_{4}T = \begin{bmatrix} c\theta_4 & -s\theta_4 & 0 & 6 \\ 0 & 0 & -1 & 0 \\ s\theta_4 & c\theta_4 & 0 & 0 \\ 0 & 0 & 0 & 1 \end{bmatrix} \tag{7.1.5}
$$

$$
{}^{5}_{4}T = \begin{bmatrix} c\theta_3 & -s\theta_3 & 0 & 6 \\ 0 & 0 & 1 & 0 \\ -s\theta_3 & -c\theta_3 & 0 & 0 \\ 0 & 0 & 0 & 1 \end{bmatrix} \tag{7.1.6}
$$

$$
{}^{6}_{5}T = \begin{bmatrix} c\theta_4 & -s\theta_4 & 0 & 6 \\ 0 & 0 & -1 & 0 \\ s\theta_4 & c\theta_4 & 0 & 0 \\ 0 & 0 & 0 & 1 \end{bmatrix} \tag{7.1.7}
$$

In this case, the former coordinate systems of the operable head move backward and new coordinate systems are produced as the robot advances by ΔL every time interval. The transformation matrix between these coordinate systems can be calculated in turn, as well as the coordinates of these points.

The foundational principle of this reconstruction algorithm entails segmenting the tendon-sheath into discrete components, each analogized as an individual joint. Subsequently, by systematically connecting these delineated points, the algorithm facilitates the simulation of the articulated posture of the robot. Finally, a fitting curve for these points is obtained, which can be interpreted as the configuration of the flexible body along with the outer sheath.

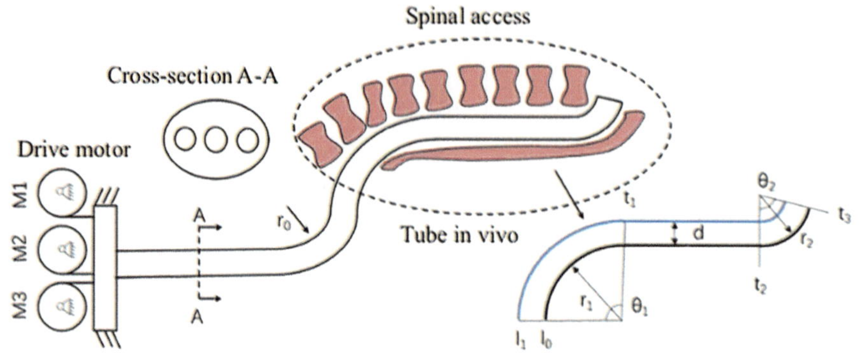

Fig. 7.2 Bending pattern perception analysis of multi-cord drive mechanism

7.2 Body Shape Reconstruction from Near-End Sensing

For the case that the slender robot cannot install sensors due to narrow or complex internal environments, or disinfection requirements, we can use the external sensor at the near end to estimate the posture and position of the robot. As shown in Fig. 7.2, a slender robot is used to pass through the natural passages of human body. We propose a method to reconstruct the bending shape of the robot inside the body using its length and shape outside. Since the cables in the robot are arranged in parallel, if the robot bends, any bending will cause variations in cable lengths at different radial positions in the sheath. As shown in Fig. 7.2, assuming that the cables are spaced at a distance d in the sheath, the following geometric relationship can be obtained:

$$\begin{cases} \theta_1 r_1 = L_0\,(\mathrm{t}_1) \\ \theta_1\,(r_1 + d) = L_1\,(\mathrm{t}_1) \end{cases} \tag{7.2.1}$$

Thus, if $L_1\,(\mathrm{t}_1)$ and $L_0\,(\mathrm{t}_1)$ are known, it follows that θ_1 and r_1.

The length of the center cable is equal to the sheath length, i.e. the cable driven by M_2 motor in Fig. 7.2 is cable L_0. The inside sheath length $L_0\,(\mathrm{t}_1)$ = total sheath length—outside sheath length. Thus, if we know the total sheath length and how each cable moves, we can determine the inner sheath length and its bending changes. This can be done by measuring the sheath length and its bending outside the body. Additionally, we can measure the external shape using vision or optical fiber sensing.

We assume that as the robot moves inside the body, its path remains consistent. This means each new body segment maintains the bend created by the previous one. Throughout this movement, the cables are always tensioned, and we overlook any minor deformations.

Figure 7.3 shows the displacement changes from its starting point, according to which the direction of movement of the casing can be known. This diagram helps us determine the robot's bending direction. If the cable movement is more pronounced

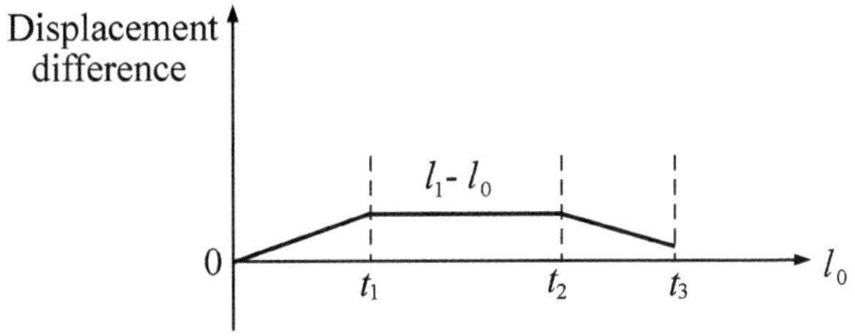

Fig. 7.3 Plot of the differential change in displacement of the cables relative to the center cable

than initially, it indicates the robot is bending to the right. On this basis, we can continually track and reconstruct the robot shape as it enters the body.

This method assumes the robot's path stays consistent as it moves inside the body. However, natural body movements or changes in pressure can disturb this path. Therefore, it's essential to conduct experiments to evaluate the performance of shape reconstruction and explore methods to improve the accuracy.

7.3 Body Shape Reconstruction from Distal-End Sensing

The shape reconstruction of a slender robot usually involves measuring the curvature at different points and then determining its 3D coordinates and overall form. However, for slender robots where it's challenging to attach strain sensors or fiber optic gratings, we proposed a different method that uses only the kinematics and sensing data of the robot head joint to estimate its shape and end position [3]. The detailed steps are outlined below.

1. Firstly, establish the configuration of the robot head joints. Create the base and end coordinate systems based on the arrangement of the head joints. If the head joints have a total of j degrees of freedom, we can estimate the body shape using based on the kinematics. This involves measuring the motion change of each joint over a specific time, as described in the subsequent steps.

$$g_{st}(\boldsymbol{\theta}) = \left(\prod_{i=1}^{j} e^{\hat{\xi}_i \theta_i}\right) g_{st}(0) \tag{7.3.1}$$

2. The robot moves forward in equal steps. Each time it advances, the total length of the head joints determines the step size. After each step, the angle of each joint is measured and recorded. As the robot progresses through n steps to reach its target position, its shape reflects the positions of the head joints at each stage. We can approximate the robot shape using the following equation.

$$g_{st}(\boldsymbol{\theta}) = \left[\prod_{k=1}^{n}\left(\prod_{i=1}^{j} e^{\hat{\xi}_i \theta_i (t_k)}\right)\right] g_{st}(0) \tag{7.3.2}$$

Where $\hat{\boldsymbol{\xi}}_i$ is the amount of joint motion rotation at i and $\theta_i\ (t_k)$ is the angular value of the i joint at at the j step.

Therefore, a 3D reconstruction method of the slender robot with the multi-joint manipulator is proposed. This method determines the configuration of the outer sheath based on the rotation angles of the four joints in the steering head. Initially, we assume that both the steering head and the flexible body maintain consistent postures as they move through any point along the robot's path. As shown in Fig. 7.4, the D-H (Denavit-Hartenberg) method was utilized to establish the spatial coordinate system for steering head. The length of the link rods between these joints are L_1, L_2, L_3, and L_4, which can be measured in the steering head; the rotation angles of the four joints are θ_1, θ_2, θ_3, and θ_4; A, B, and C are three reference points. The transformation matrix between each coordinate system can be calculated as follows

$$ {}^{0}_{1}T = \begin{bmatrix} c\theta_1 & -s\theta_1 & 0 & 0 \\ s\theta_1 & c\theta_1 & 0 & 0 \\ 0 & 0 & 1 & 0 \\ 0 & 0 & 0 & 1 \end{bmatrix} \tag{7.3.3} $$

$$ {}^{l}_{2}T = \begin{bmatrix} c\theta_2 & -s\theta_2 & 0 & L_1 \\ 0 & 0 & -1 & 0 \\ s\theta_2 & c\theta_2 & 0 & 0 \\ 0 & 0 & 0 & 1 \end{bmatrix} \tag{7.3.4} $$

$$ {}^{2}_{3}T = \begin{bmatrix} c\theta_3 & -s\theta_3 & 0 & L_2 \\ 0 & 0 & 1 & 0 \\ -s\theta_3 & -c\theta_3 & 0 & 0 \\ 0 & 0 & 0 & 1 \end{bmatrix} \tag{7.3.5} $$

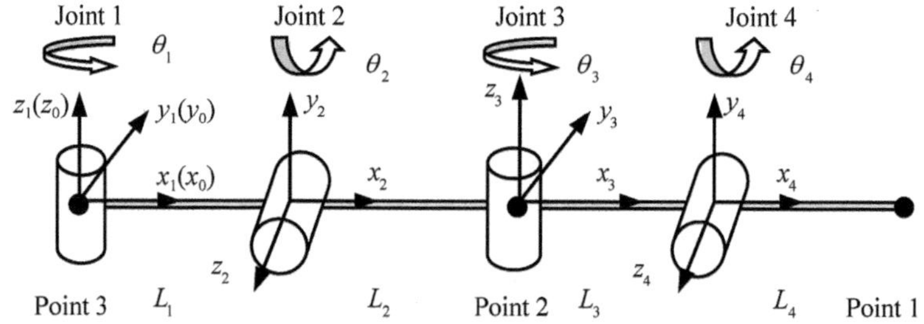

Fig. 7.4 Coordinate system of the steering head (Reprinted from [4] in Chap. 5 with permission from IEEE)

$$
{}^{3}_{4}T = \begin{bmatrix} c\theta_4 & -s\theta_4 & 0 & L_3 \\ 0 & 0 & -1 & 0 \\ s\theta_4 & c\theta_4 & 0 & 0 \\ 0 & 0 & 0 & 1 \end{bmatrix} \tag{7.3.6}
$$

If Point 3 is the coordinate origin at the initial moment k, transform matrixes of Point 1 and Point 2 are acquired as follows.

$$
{}^{0}_{4}T(k) = {}^{0}_{1}T(k) \cdot {}^{1}_{2}T(k) \cdot {}^{2}_{3}T(k) \cdot {}^{3}_{4}T(k) \tag{7.3.7}
$$

$$
{}^{0}_{2}T(k) = {}^{0}_{1}T(k) \cdot {}^{1}_{2}T(k) \tag{7.3.8}
$$

Such that at the initial moment k, the coordinates of point 2 (P_2) and point 1 (P_1) are

$$
P_2(k) = {}^{0}_{2}T(k) \cdot P_3(k) \tag{7.3.9}
$$

$$
P_1(k) = {}^{0}_{4}T(k) \cdot P_3(k) \tag{7.3.10}
$$

Then at the following moments $k+1$ and $k+2$, the tendon-sheath robot moves forward $\Delta L(\Delta L = L_1 + L_2 + L_3 + L_4)$ twice. The coordinates of former points change correspondingly and new points at former coordinate system are produced.

$$
k+1 \longrightarrow \begin{cases} P_4(k+1) = P_2(k) \\ P_3(k+1) = P_1(k) \\ P_2(k+1) = {}^{0}_{2}T(k+1) \cdot P_1(k) \\ P_1(k+1) = {}^{0}_{4}T(k+1) \cdot P_1(k) \end{cases} \tag{7.3.11}
$$

$$
k+2 \longrightarrow \begin{cases} P_6(k+2) = P_4(k+1) \\ P_5(k+2) = P_3(k+1) \\ P_4(k+2) = P_2(k+1) \\ P_3(k+2) = P_1(k+1) \\ P_2(k+2) = {}^{0}_{2}T(k+2) \cdot P_1(k+1) \\ P_1(k+2) = {}^{0}_{4}T(k+2) \cdot P_1(k+1) \end{cases} \tag{7.3.12}
$$

In this case, the former coordinate systems of the operable head move backward and new coordinate systems are produced as the robot advances by ΔL during each interval. The transformation matrix between these coordinate systems can be calculated in turn, as well as the coordinates of these points.

Essentially, this algorithm of reconstruction is to divide the tendon-sheath into smaller segment and each segment is regarded as a joint. By connecting these calculated points, we can caculate the robot posture. Finally, the fitting curve of these points are obtained, which can be seen as the configuration of the flexible body as well as the outer sheath.

In order to verify the practical effect of the method, the experimental pipes shown in Fig. 7.5 were constructed to reconstruct the shape of the robot motion in the plane and the motion process in space, respectively. A hollow circular pipe with a diameter of 60 mm is bent into a semicircle with a diameter of 0.8 m in Fig. 7.5a. The robot shape in real time is shown in Fig. 7.6, and its reconstructed shape and the actual shape of the bent tube are compared in Fig. 7.7. The shape reconstruction display interface (Fig. 7.6) was created by C++ Builder and OpenGL modules. The scaled virtual shapes are drawn in the main window according to the actual size of the slender robot. It can be seen that the reconstructed shape is basically the same as the actual shape of the circular pipe, but it does not coincide exactly with the centerline of the circular pipe. This discrepancy arises because there's a gap between the robot and the tube. Additionally, the flexible body has inherent bending stiffness, causing it to adhere to the concave side of the inner wall of the tube as it moves forward. In Fig. 7.5b, the circular tube is bent into two $\frac{1}{4}$ arcs of 0.4 m radius, and the two arcs are located in two mutually perpendicular planes. The shape reconstruction when the robot moves inside this circular tube is shown in Fig. 7.8, which basically reflects the actual shape of the circular tube. The comparison of the two shapes is shown in Fig. 7.9, and as in the case of the plane the robot is also moving closely to the concave side of the inner wall of the circular tube.

Localization of the target can also be achieved using the shape reconstruction method. The data in Tables 7.2 and 7.3 show the comparison between the coordinates of the robot head and the circular tube. The average values of the spatial coordinates of the robot end are obtained from several experiments, and the difference between the average value of the end coordinates and the actual coordinates of the end of the circular tube is taken as the positioning error value. It can be seen that the maximum error of target localization using the kinematic approximation of the robot head joints is only 6.17 cm, which basically meets the requirements of the robot for search and rescue detection in confined spaces.

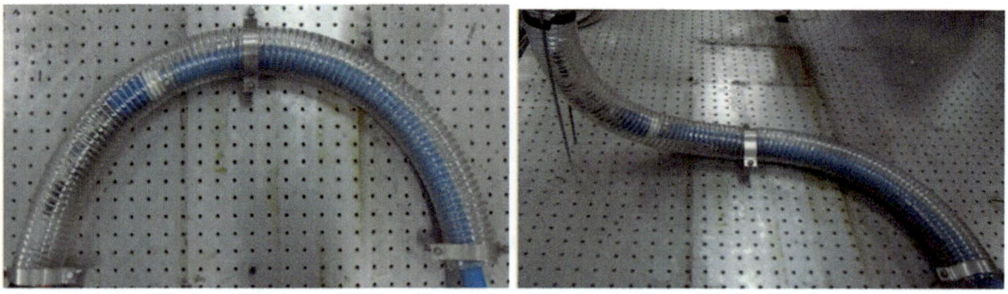

(a) (b)

Fig. 7.5 Comparison of actual shape and fit shape calculated by the 3D reconstruction method, where (**a**) and (**b**) show two bending shapes tested in the experiment [4]

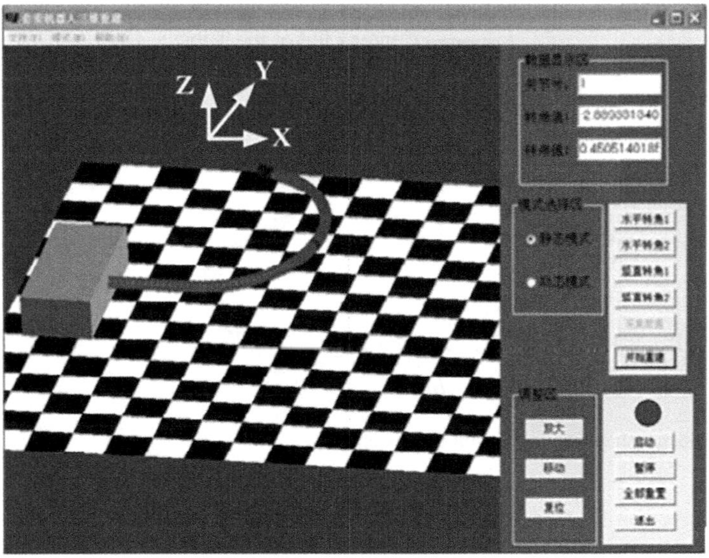

Fig. 7.6 Planar shape reconstruction display effect [4]

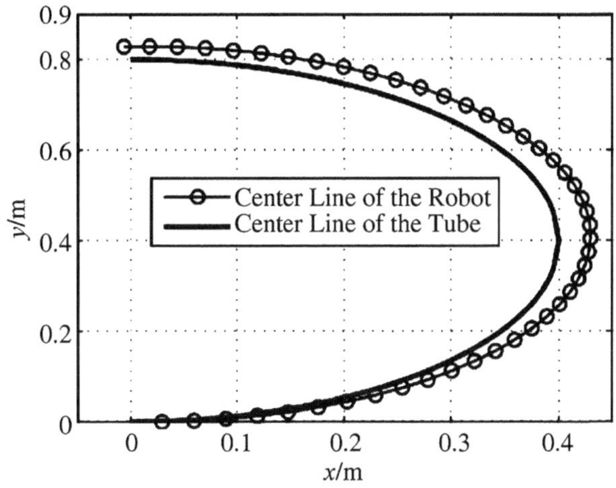

Fig. 7.7 Comparison of plane shape reconstruction effect [4]

7.4 Geometry Shape Sensing of Sheath Configuration

From the transmission model of tendon-sheath mechanism, we can find that the friction of tendon-sheath mechanisms is heavily dependent on the sheath configuration. However, the static model used for compensation control cannot deal with the variation of sheath path. To enhance the control accuracy of tendon-sheath mechanism

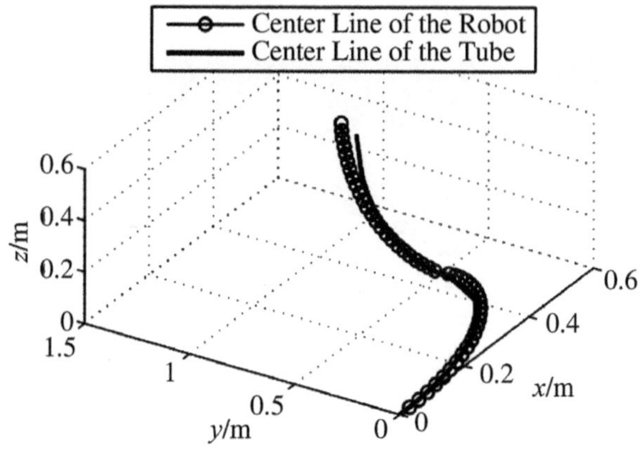

Fig. 7.8 Spatial shape reconstruction display effect [4]

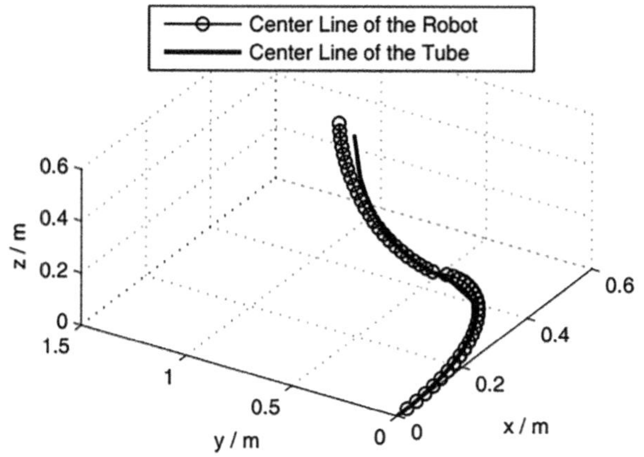

Fig. 7.9 Comparison of spatial shape reconstruction effect [4]

Table 7.2 Plane shape positioning error values/cm

	x axis coordinates	y axis coordinates
Actual spatial coordinate value	0	80
Reconstruction of spatial coordinate values	0.62	82.8
Positioning error value	0.62	2.8

with time-varying configuration, a tendon-sheath-based bending sensor is proposed for the online identification of accumulated bending angle [5].

As depicted in Fig. 7.10, for the outer sheath made up of tightly helical coils, the tendon line passing through the outer sheath varies with the system configuration.

Table 7.3 Spatial shape positioning error values/cm

	x axis coordinates	y axis coordinates	z axis coordinates
Actual spatial coordinate value	45	90	45
Reconstruction of spatial coordinate values	42.3	93.06	51.17
Positioning error value	−2.7	3.06	6.17

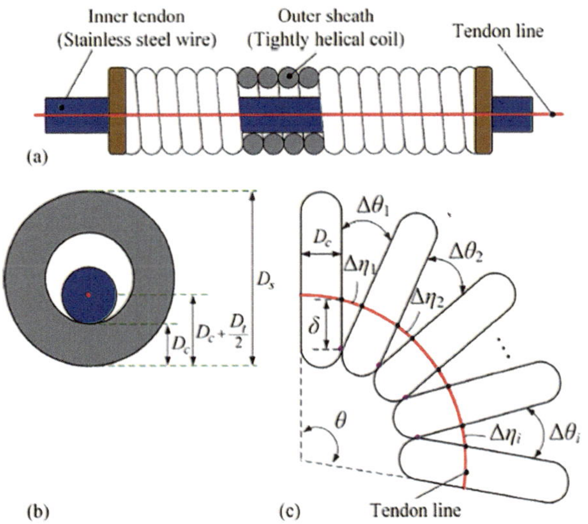

Fig. 7.10 Schematic diagram of the deformation of outer sheath with different configuration (Reprinted from [5] with permission from Elsevier)

Jeong et al. [6] proposed a model to reveal the relationship between accumulated bending angle and the central displacement change of coil. The model is developed based on the simplification of the geometric model of helical coil and the Taylor series approximation. Besides, the bending angle of different coils is assumed to be the same, which is unsuitable for practical application and leads to modeling error. Thus, an optimized model is proposed in this section. As can be observed in Fig. 7.10c, the actual geometric model of helical coil has a cross-sectional shape combined with square and circular. The displacement of the tendon line increases with the bending angle of tendon-sheath mechanism, since the gap between each helical coil lead to the equivalent elongation of outer sheath. The cross section of the tendon sheath component is depicted in Fig. 7.10b, where D_s denotes the external diameter of the helical coil; D_c represents the diameter of the steel wire used to make helical coil; D_t denotes the diameter of the inner tendon. It can be seen that there exist gap between the outer sheath and inner tendon, and it can effectively

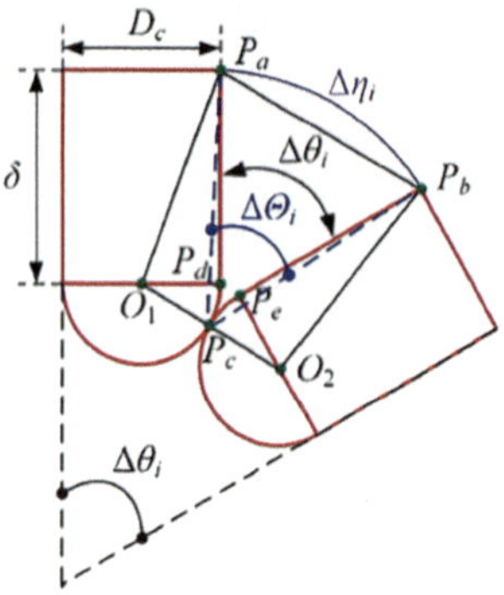

Fig. 7.11 Schematic diagram of the contact point and bending angle between two adjacent helical coils (Reprinted from [5] with permission from Elsevier)

prevent excessive friction. Therefore, the tendon line and the center of sheath are misaligned in practical operation. The inner tendon with external tension contacts the inner surface of helical coils when the sheath bends, as shown in Fig. 7.10b.

The relation among the bending angle, the deformation of outer sheath, and the displacement change of inner tendon is shown in Fig. 7.10c. The helical coils keep in contact at one end as the increase of bending angle, and the accumulated bending angle of outer sheath can be expressed as follows:

$$\theta(t) = \sum_{i=1}^{n-1} \Delta\theta_i(t) \tag{7.4.1}$$

where $\Delta\theta_i(t)$ represents the angle between two adjacent helical coils. n denotes the total number of the helical coils.

Figure 7.11 shows the close view of the contact point and bending angle between two adjacent helical coils. The distance between the contact point of each coil and the tendon line increases with the bending angle. In the close view diagram, O_1 and O_2 denote the centers of two adjacent helical coils on the same cross section. P_c is the contact point located in the line between O_1 and O_2. The distance between the tendon line and the center of helical coil is denoted as $\delta = (D_c + D_t)/2$. Besides, P_a, P_b, P_d, P_e represent four feature points on the cross section of helical coils, which satisfy: $|P_a P_d| = |P_b P_e| = \delta$, $|O_1 P_d| = |O_2 P_e| = |O_1 O_2|/2 = D_c/2$. If $\Delta\theta_i(t) = 0$, points P_a and P_d are coincided with points P_b and P_e respectively. Then, with the increase of bending angle, the contact point of helical coil will leave from point $P_d(P_e)$ to point P_c on the arcs of helical coils. Therefore, the displacement change of inner tendon between two adjacent helical coils can be presented as the arc length $\Delta\eta_i$ with $|P_a P_c|$ as radius, point P_c as center, and $\Delta\Theta_i(t)$ as central angle.

According to the geometric relations shown in Fig. 7.11 and the Law of Cosines, we have

$$
\begin{aligned}
|P_aP_c| &= \sqrt{|P_aO_1|^2 + |P_cO_1|^2 - 2\cos(\angle P_aO_1P_c)\cdot|P_aO_1|\cdot|P_cO_1|}\\
&= \sqrt{\frac{2\delta^2 + D_c^2}{2} - \sqrt{\frac{4D_c^2\delta^2 + D_c^4}{4}}\cos\left[\frac{\pi + \Delta\theta_i(t)}{2} - \arctan\left(\frac{D_c}{2\delta}\right)\right]}
\end{aligned}
\tag{7.4.2}
$$

$$
\Delta\Theta_i(t) = \pi - 2\arccos\sqrt{\frac{|O_1P_c|^2 + |P_aP_c|^2 - |O_1P_a|^2}{2|O_1P_c|\cdot|P_aP_c|}} = \pi - 2\arccos\sqrt{\frac{|P_aP_c|^2 - \delta^2}{D_c\cdot|P_aP_c|}}
\tag{7.4.3}
$$

Thus, the theoretical arc length $\Delta\eta_i$ can be expressed a function of $\Delta\Theta_i(t)$ and calculated as follows:

$$
\Delta\eta_i = |P_aP_c|\cdot\Delta\Theta_i(t) \tag{7.4.4}
$$

We can find that the relations among $\Delta\theta_i$, $|P_aP_c|$ and $\Delta\Theta_i(t)$ are apparently nonlinear. The accurate radius and central angle are difficult to be directly obtained from measurement. From Fig. 7.11, it can be observed that in the original straight state (i.e., $\Delta\theta_i = 0$), the radius of arc $\Delta\eta_i$ is equal to δ, and the central angle $\Delta\Theta_i(t)$ is equal to the bending angle $\Delta\theta_i(t)$. The difference between $|P_aP_d|$ and δ, as well as that between $\Delta\Theta_t(t)$ and $\Delta\theta_i(t)$, is small with a small bending angle between two adjacent helical coils. To simplify the calculation of arc length $\Delta\eta_i$, the radius is approximated to be a constant (i.e., δ), and the central angle is assumed to be equal to the bending angle if the bending angle is small enough. The estimated arc length $\Delta\tilde{\eta}_i$ can be expressed as follows:

$$
\Delta\tilde{\eta}_\mathrm{i} = \delta\cdot\Delta\theta_i(t) \tag{7.4.5}
$$

The theoretical arc length based on Eq. (7.4.4) and the estimated arc length based on Eq. (7.4.5) with different bending angle between two adjacent helical coils are compared. The parameters of tendon sheath are set as: D_s = 3 mm, D_t = 1 mm, D_c = 0.5 mm. It indicates that the estimated error is increase with the bending angle. Furthermore, the estimated error can be limited within 0.8% if $\Delta\theta_i(t)$ is less than 5 °C. In the practical application of tendon-sheath mechanism, the average $\Delta\theta_i(t)$ is usually limited within 2 °C. Thus, the estimated error caused by this assumption is negligible.

Thus, from Eqs. (7.4.1) and (7.4.5), the overall displacement change of inner tendon with different bending angle can be obtained and described as follows:

$$
\Delta\eta(t) \approx \Delta\tilde{\eta}(t) = \sum_{i=1}^{n-1}\delta\cdot\Delta\theta_i(t) = \delta\cdot\theta(t) = \frac{D_c + D_t}{2}\theta(t) \tag{7.4.6}
$$

where $\Delta\eta(t)$ and $\Delta\tilde{\eta}(t)$ represent the theoretical and estimated displacement change of the entire inner tendon resulting from the deformation of outer sheath. It can be

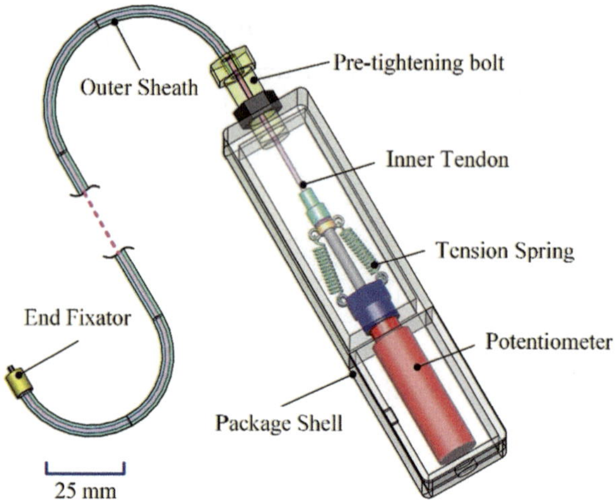

Fig. 7.12 Schematic of the tendon-sheath-based bending sensor (Reprinted from [5] with permission from Elsevier)

observed that the displacement change of inner tendon is approximately linear to the accumulated bending angle.

To achieve real-time identification of the tendon-sheath mechanism configuration, we developed a bending sensor. This sensor detects the accumulated bending angle of an outer sheath, considering its significant curvature and changing shape over time. This detection relies on the relationship between the displacement change of the inner tendon and the bending angle, as described in Eq. (7.4.6). The major components of the bending sensor are shown in Fig. 7.12. The bending section consists of a sheath made of helical coils, with an outer diameter of 3 mm and an inner diameter of 2 mm. Inside this sheath is a stainless steel wire with a diameter of 1 mm. The distal end of the inner tendon connects to the sheath end, while its proximal end links to a high-precision linear potentiometer (FIAYE, LSM-20, Resolution: 1 lm, Linearity: 0.1%). This potentiometer measures the displacement of the inner tendon due to changes in the bending angle. The proximal end of sheath is fixed to a pre-tightening bolt utilized to modulate the system pretension and avoid slacking problem. Two tension springs are integrated into the sensing part to keep the tendon under tension state.

The usage of the proposed bending sensor is simple and reliable. The flexible bending part (sensing wire) of bending sensor needs to be connected in parallel to the outer sheath of actuation tendon-sheath mechanism via several sheath fixators. The configuration of bending sensor is the same as the configuration of tendon-sheath mechanism. Then, the sensing part is located to the actuation side to measure the displacement variation. Since the electronic components of the sensors are placed only at the proximal end, it is practical to the applications with severe working conditions (e.g., oil stain, spray water, severe dust) or strict dimension limitation (e.g., surgical robot and endoscopic system). Evaluation experiments were carried out to analyze the accuracy of the proposed bending sensor with a discrete bending angle that changes from 0° to 540° with an interval of 45°. The experimental results are shown in Fig. 7.13. It can be seen that the measured bending angle is basically consistent with the reference bending angle, with a root mean square error (RMSE) of 7.60° and a mean absolute percentage error (MAPE) of 3.9%. The measured error is mainly caused by the deformation of sensing tendon and the estimated error of displacement variation in Eq. (7.4.6). Nevertheless, the proposed bending sensor has positive effect on improving the force and position control performance of the tendon-sheath mechanism without feedback from the distal end. Finally, it should be pointed out that the bending sensor is capable of measuring the accumulated bending angle of the tendon-sheath mechanism. The configuration of tendon-sheath mechanism cannot be determined via the bending sensor, as there could be many sheath configurations corresponding to the same accumulated bending angle.

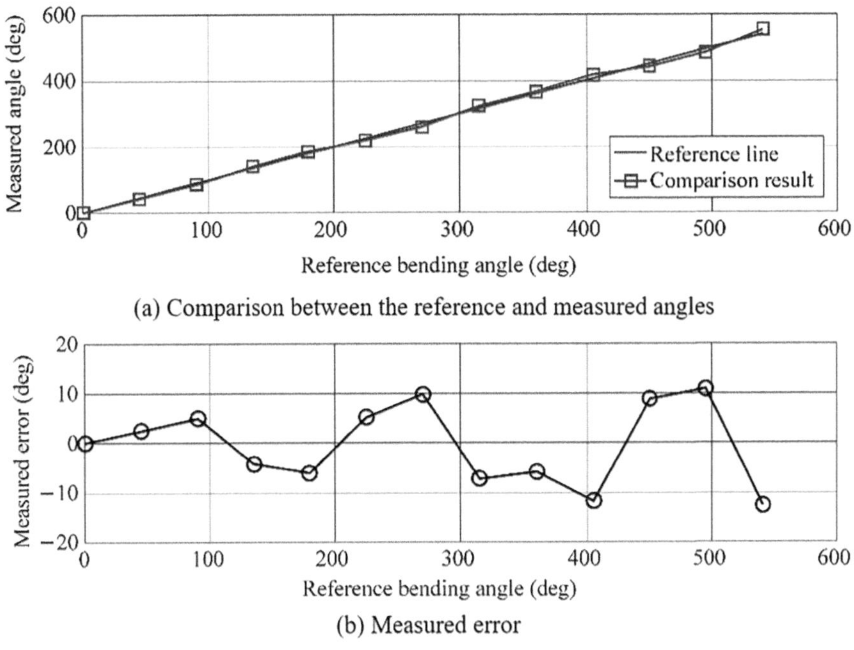

Fig. 7.13 Experimental result with a discrete accumulated bending angle that changes from 0° to 540° with an interval of 45° (Reprinted from [5] with permission from Elsevier)

7.5 Fiber Bragg Grating for Sensing Continuum Body Shape

Optical fiber sensing technology uses optical fibers to detect changes in external environmental factors. Optical fiber, also known as optical waveguide fibers, as a cylindrical waveguide structure based on the principles of total internal reflection, typically constructed from materials such as quartz glass. Optical fibers consist of three concentric layers: core, cladding, and protective coating. The core, enriched with a certain concentration of impurities to increase its refractive index, forms the innermost layer. The cladding, with a slightly lower refractive index than the core, serves to facilitate total internal reflection at the core-cladding interface. The outer diameter of optical fibers typically measures 125 m, with an additional coating layer of epoxy resin or silicone gel, ranging from 5 to 40 m in thickness, applied for protection against contamination and physical damage.

The inception of optical fiber optics can be traced back to the formation of optical fiber waveguide structures in the mid-twentieth century. Since the pioneering work of Chinese-American scientist Kao Kuen and his colleagues, who first reported in 1966 that optical fibers could be employed as a medium for long-distance signal transmission, the field of optical fiber sensing technology has experienced rapid development. It has given rise to a diverse range of optoelectronic devices [3].

Optical fiber sensing can be classified into two major categories: transmission-type and sensing-type. The former primarily employs optical fibers as a means of transmitting information and demands that optical fibers possess long-distance, high-capacity, and low-loss transmission characteristics to ensure accurate feedback of sensor data to the signal processing system. The latter directly employs optical fibers themselves as sensing elements, achieving sensing by monitoring changes in parameters such as optical intensity and phase during variations in the external environment. Simultaneously, it utilizes optical fibers for data transmission, hence referred to as functional sensing. Fiber Bragg Gratings (FBG) is a main research area of the functional sensing. FBG exhibit excellent temperature and strain responsiveness, suitable for both point and quasi-distributed sensing systems.

Over nearly four decades of continuous progress and development, FBG evolved into crucial passive optical devices in fields such as fiber-optic communications, fiber-optic sensing, and optical information processing. As of now, FBG sensing has been successful in measuring a wide range of physical quantities, including temperature, strain, refractive index, and vibration. Compared with other sensors, fibre optic grating sensors have the advantages of small size, light weight, anti-electromagnetic interference, easy to reuse and so on. Fibre grating can be used as a quasi-distributed sensor in the form of a sensing network to monitor a wide range of areas, and can also be used as a distributed sensor to sense a small range of strain gradients, and then monitor the crack extension.

With the development of optical fiber communications and sensing technology, optical fiber grating sensing technology in traditional fields such as aerospace, petroleum engineering, civil engineering and construction have more mature applications. With the higher expectations and demands of doctors and patients for minimally invasive medical treatment, fibre grating sensing has become a hot spot for minimally invasive medical sensing in recent years. Understanding the level and current status of minimally invasive medical FBG sensing technology and recognising its advantages and shortcomings can promote the higher and faster development of FBG sensing technology in the field of minimally invasive medical treatment and improve the safety, quality and efficiency of minimally invasive medical treatment.

7.5.1 Fibre Bragg Grating Sensing Principle

FBG is a wavelength-modulated sensing element, the grating is engraved into the core of the fibre in a periodic arrangement, the wavelength of the incident light has a selective effect. According to the coupling theory, when a beam of broadband light source into the fibre grating, due to the different refractive indices of the optical fibre will occur reflection and transmission phenomena, only to meet the specific wavelength conditions of the light signal will be reflected back, the rest of the light signal will be transmitted, thus creating a depression in the transmission spectrum, in the reflection spectrum presents a wave peak. Fibre bragg grating sensing principle is that when the fibre grating is subjected to external forces and temperature, the reflected wave peak will be shifted, that is, the Bragg wavelength will be changed, through the decoupling of the wavelength change we can obtain the change of the external environment.

The reflected Bragg wavelengths need to satisfy the following Eq. 7.5.1,

$$\lambda_B = 2 \text{ neff } \Lambda \tag{7.5.1}$$

where: neff is the effective refractive index of the fibre core; Λ is the gratisng period. From the formula (7.5.1) can be seen, Bragg wavelength λ_B is determined by the effective refractive index of the fiber core neff and grating period Λ, and varies with their changes. When the fibre Bragg grating is subjected to external stress and temperature, by the elastic optical effect and thermo-optic effect, the effective refractive index of the fibre grating and grating period will change, so the reflected Bragg wavelength λ_B of the change in the formula can be expressed as follows

$$\Delta\lambda_B = K\varepsilon\Delta\varepsilon + KT\Delta T \tag{7.5.2}$$

where: $\Delta\lambda_B$ is the central wavelength offset; $K\varepsilon = 1 - p_e$ is the wavelength strain sensitivity coefficient under stress, where p_e is the effective elastic-optical coefficient; $KT = \alpha + \xi$ is the sensitivity coefficient for temperature sensing of the optical fibre grating, with α being the coefficient of thermal expansion and ξ being the thermo-optical coefficient; $\Delta\varepsilon$ is the axial strain variation; ΔT is the temperature variation.

Using the most commonly used pure fused silica, the general values of the parameters are $p_e = 0.216\varepsilon = -1, \alpha = 5.510 - 7C - 1, \xi = 6.310 - 61C - 1$.

7.5.2 Current Research on FBG Sensors for Shape Reconstruction

FBG sensors have the advantages of small size and rapid signal transmission, making them ideal for real-time shape detection in minimally invasive surgical devices. Researchers have achieved stage-by-stage results in using FBG sensors to reconstruct the shape of minimally invasive surgical devices and proposed different methods. Zhang et al. integrated two FBG sensors into a colonoscope. One sensor was used for progressive detection and the other for real-time dynamic reconstruction. Moon et al. integrated three optical fibres in a minimally invasive surgical device, each housing four FBG sensors operating at varied wavelengths. This setup allows simultaneous detection of curvature and torsion, ensuring precise 3D shape reconstruction. Additionally, three FBGs at the same position enable temperature compensation [7]. Park et al. explored integrating three optical fibers with two FBG groups into a flexible needle to determine its shape within the human body. Although they achieved high accuracy,their approach primarily addressed needle deflection in free space and only considered curvature at two points along its axis. Berg et al. proposed a design for an actively controlled manipulable needle with tendon actuation and FBG-based shape sensing. Xu et al. utilized a helically-wound FBG sensor to measure curvature, torsion, and force in a continuum-type robot [8].

For structures employing more than three fibres, Floris et al. proposed a new method for twist compensation in optical shape sensing based on the St. Venant torsion theory of homogeneous cylinders. Their method employs FBGs inscribed in spun multi-core fibers (MCFs) for precise twist sensing. They refined their technique to calculate twist and reconstruct 3D shapes, compensating for detected twists [9]. Similarly, Khan et al. used FBG-inscribed multicore fibers to measure the orientation of segmented conduits. By converting sensor readings into strains and using specific equations, they reconstructed the conduit shape. Lu et al. used FBG sensors for 3D shape sensing in flexible endoscopy, proposing a method to determine the endoscope's curvature and distortion in three dimensions [10].

Finally for the study of the nature of fibre optic sensors, Floris et al. studied the performance of MCF and FBG sensors for shape sensing, finding that longer FBG sensors offer better accuracy. Lezcano et al. proposed the trajectory generation of FBG sensing needles for insertion into multilayered tissues, and proposed a method to generate the trajectory of flexible needles. Yao et al. proposed the design and analysis of a combined FBG sensor with three-parameter measurements, and designed an FBG sensor based on the FBG sensor to generate a flexible needle.

7.6 Force and Displacement Sensing of Micro Tip Manipulation Using FBG Sensors

Force sensing in continuum robots has been studied a lot, where the applied force is estimated from tip deflection or joint force. These works show that the sensing capabilities of continuum robots are highly dependent on the inherent compliance of the robot body. For robots that are laterally flexible and axially incompressible, it is difficult to have high quality force measurements in all three dimensions. FBG based 3D tip force sensors have been developed for catheter ablation and vitreoretinal surgery by modifying the axial stiffness of the sensing structures; however, these structures do not have a hollow center for passing micro-surgical instruments or delivering therapy. A lateral force sensing system based on FBG sensors was developed for a continuum robot with a hollow center, but the method proposed requires that the sensing structure be initially straight. We focuses on developing high accuracy lateral force measurement on precurved continuum robots, embedding FBG sensors into the robot body to provide dynamic force measurement and protect the fibers.

To address the challenge of simultaneously measuring curvature, torsion and applied lateral force on straight and precurved continuum robots, the design and manufacture of a novel helically-wrapped FBG sensing technology is proposed and a force-curvature-strain model is developed. The helical grooves on a super-elastic cylindrical structure is created by a high-precision custom engraving technique. To measure the curvature and torsion in the robot from the strain in the helically-wrapped fibers a curvature-strain model is developed. Furthermore, a force curvature-strain model is derived, incorporating a nonlinear mapping between the tip force and strain in the FBG sensors. The calculation of the inverse of this model is then proposed to obtain accurate force measurements from the FBG readings with minimal computational cost. The novel sensing technology and calculation techniques proposed are evaluated experimentally by embedding three FBG sensors into a precurved Nitinol tube at the same cross-section. The results demonstrate that this technology can provide accurate and sensitive curvature, torsion and force measurement for continuum robots.

The performance of the FBG depends on its reflection spectrum, which is decoded using an incoming light source. This spectrum helps detect changes in physical parameters such as strain and temperature by noting shifts in the central reflected wavelength. The central wavelength of the reflected light can be defined as Eq. 7.5.1. Any external disturbance resulting in a change in temperature or axial strain within the fiber will cause an analogous shift in the reflected central wavelength. The magnitude of this shift corresponds with the magnitude of the disturbance; and as such, the dependence of the central wavelength shift on temperature and strain can be described as:

$$\frac{\Delta\lambda_B}{\lambda_B} = (1 - \rho_e)\,\Delta\varepsilon + (\alpha_f + \xi_f)\,\Delta T \tag{7.6.1}$$

where. p_e is the photo-elastic coefficient of the fiber core; a f. α_f, ξ_f are the thermal expansion and thermo-optic coefficient of the optical fiber. $\Delta\varepsilon$ is the variation of the longitudinal strain and ΔTis the temperature change.

With the improvement of medical equipment and technology, various tissue ablation procedures are widely used for the treatment of cardiac diseases and tumours. In the absence of force feedback between the end of the ablation catheter and the tissue, surgeons can only assess the catheter-tissue contact status through medical imaging or other indirect methods, so the acquisition of force feedback information during tissue palpation has been a hot topic in recent years.

In 2020, Shuxin Wang's team at Tianjin University designed a FBG palpation force sensor to achieve axial force sensing of slender surgical instruments in minimally invasive medical treatment and to determine tissue abnormalities in minimally invasive surgery [11]. The force-sensing curved portion of this palpation catheter has four curved hinges uniformly arranged at 90° intervals along its circumference to form a symmetric structure, and the optical fibre is tightly suspended in a small groove in the force-sensitive curved mid-axis, which avoids cross-interference and chirp failure from radial forces. The palpation experiments show that the flexible design based on the Sarrus mechanism achieves a good axial linear force-deformation relationship by using the method of rigid body replacement, and achieves a high resolution of 2. 55 mN in a relatively large measurement range of 0–5 N.

Shuxin Wang's team achieved a significant breakthrough in structural design. Their flexible force-sensing bending part, based on the Sarrus mechanism, exhibits an excellent linear relationship between force and deformation, along with an expanded measurement range. Tianliang Li's team has done a series of studies on tissue palpation and also designed different novel structures to achieve the improvement from discrete palpation to continuous sliding palpation, which enhances the efficiency of palpation, meanwhile, a method to decouple the triaxial force component and temperature is significant in his research. Although the existing tissue palpation in various studies has not yet reached the practical application in medicine, the quantitative palpation technique based on FBG has a promising development prospect.

References

1. Wang T, Lin B, Chong B et al (2021) Reconstruction of backbone curves for snake robots. IEEE Robot Autom Lett 6(2):3264–3270
2. Shi C, Luo X, Qi P et al (2016) Shape sensing techniques for continuum robots in minimally invasive surgery: a survey. IEEE Trans Biomed Eng 64(8):1665–1678
3. Li J, Wang X, Xi R (2016) Tip force tracking control based on the tendon-sheath transmission model and 3D reconstruction of a long slender rescue robot. In: 2016 IEEE 14th international workshop on advanced motion control (AMC). IEEE, pp 448–454
4. Zhang G (2014) Three dimensional reconstruction of a tendon-sheath driven long and slender robot. Southeast University
5. Wu Q, Wang X, Chen B et al (2019) Modeling, online identification, and compensation control of single tendon sheath system with time-varying configuration. Mech Syst Signal Process 130:56–73

6. Jeong U, Cho KJ (2016) A novel low-cost, large curvature bend sensor based on a Bowden-cable. Sensors 16(7):961
7. Moon H, Jeong J, Kang S et al (2014) Fiber-Bragg-grating-based ultrathin shape sensors displaying single-channel sweeping for minimally invasive surgery. Opt Lasers Eng 59:50–55
8. Xu R, Yurkewich A, Patel R (2016) Curvature, torsion, and force sensing in continuum robots using helically wrapped FBG sensors. IEEE Robot Autom Lett 1(2):1052–1059
9. Floris I, Madrigal J, Sales S et al (2020) Twisting measurement and compensation of optical shape sensor based on spun multicore fiber. Mech Syst Signal Process 140:106700
10. Khan F, Donder A, Galvan S et al (2020) Pose measurement of flexible medical instruments using fiber Bragg gratings in multi-core fiber. IEEE Sens J 20(18):10955–10962
11. Lv C, Wang S, Shi C (2020) A high-precision and miniature fiber Bragg grating-based force sensor for tissue palpation during minimally invasive surgery. Ann Biomed Eng 48:669–681

Chapter 8
Tendon-Sheath Transmitted Slender Robots for Detection and Rescue

Abstract Detection and rescue robots are an important branch of current robotic research, these robots can navigate and operate in complex situations like earthquakes and building collapses. To work in the narrow spaces after disasters, there's a growing focus on developing the moving robot with slender body and flexible manipulators. In this chapter, we'll explore the environments where these slender search and rescue robots work. Then discuss the key design elements they require and introduce real-world examples. Additionally, we'll cover the steps involved in designing and building these robots. We hope this chapter can offer readers practical insights and guidance, and enable them to create their own rescue robot designs.

8.1 System Consideration of Detection and Rescue Robots

8.1.1 Background of Search and Rescue Robots

Over the past two decades, and particularly following numerous terrorist attacks, many countries have intensified their research efforts on search and rescue robots. As a result, various robots with different driving mechanisms have been reported, which mainly include tracked robots, wheeled robots, and bionic robots [1, 2].

Tracked robots were initially developed for military applications, such as the removal of hazardous objects. Their design provides them with excellent mobility on rough and unstable surfaces, enabling them to navigate complex obstacles effectively [3, 4]. However, tracked robots come with several limitations like forward friction, energy inefficiency, and the constant need for track tensioning. Additionally, due to their larger size, they are well-suited for operating on the surface of debris but face challenges accessing narrow spaces within ruins.

Wheeled robot has the advantages of simple structure, high reliability, low cost and fast moving speed on even terrains [5, 6]. However, compared with tracked robots, they exhibit limitations in navigating obstacles within intricate and rugged disaster environments, which limits their application in many scenarios.

The design inspiration of bionic robot mainly comes from nature. Common types of bionic robots include snake-like, insect-like, and fish-like designs. Snake-like

X. Wang and Q. Zhang, *Tendon-Sheath Transmitted Light and Slender Robots*,
https://doi.org/10.1007/978-981-96-5581-6_8

robots are known for their flexibility and small size, allowing them to navigate narrow spaces effectively, making them ideal for search and rescue missions in confined areas. Meanwhile, insect-like crawling robots excel at overcoming obstacles, making them suitable for surface-level searches in ruins. The most typical snake-like robot is Unified Snake [7] developed by Cornell Wright et al., Carnegie Mellon University. Unified Snake is a hyper-redundant series articulated robot with numerous articulated joints and a modular design. Each module contains a motor, gear set and SMA driven bistable brakes and cross-module connectors. Unified Snake is very flexible, it can climb trees and move quickly in deserts, quagmire, and ruins.

Among insect-like robots, the foot robot excels in searching the surface of ruins due to its great obstacle-crossing capabilities. Kevin C. Galloway et al. from the University of Pennsylvania developed X-RHex [8], a high-speed hexapod robot. The biggest feature of the robot is that it contains six sickle legs, each of which is driven by a motor. This design allows X-RHex to navigate obstacles of a specific height while maintaining a rapid pace on uneven terrain, with a maximum speed of 1.54 m/s.

8.1.2 Design of Search and Rescue Robot System

Every year, numerous natural disasters, including earthquakes, mudslides, hurricanes occur worldwide. These disasters often causes serious casualties, fire, flood, toxic gas leakage, bacteria and radioactive material diffusion, and may also cause tsunami, landslide, collapse, ground fissure and other secondary disasters. This directly leads to a very complex post-disaster environment. Most of the search and rescue work in disaster ruins was completed by search teams and rescue dogs. However, the potential secondary disasters threatened the lives of searchers. The earthquake caused numerous buildings collapsed, and trap many survivors beneath the rubble. Additionally, some passages within the debris are so narrow that neither search teams nor rescue dogs can navigate them. The characteristics of these passage mainly include the following aspects:

1. The surrounding environment of the passage is irregular and formed by collapsed buildings of different shapes and sizes.
2. There may be several meters of altitude difference between the passage and the survivors.
3. The size of the narrow passage changes in a large range.
4. The passage leading to the survivors is rough and there may be many curves of different sizes.

The slender soft robot is mainly used for searching and rescuing in these narrow spaces. These tasks involve navigating small and complex areas, so the robot design must meet specific requirements:

1. Size is important. A study by Japanese researchers found that debris suitable for post-earthquake search and rescue is typically around 10 cm in size. Taking this

into account, we designed our robot with a 40 mm diameter to navigate debris after earthquakes.
2. The robot needs good guidance. It should easily move around obstacles in tight spaces and have flexibility in its head movement.
3. The robot must adapt easily. It operates in varied environments, so it needs to adjust without detailed planning. Making the entire robot length adjustable might complicate its controls and hinder its performance in real-world scenarios.
4. Retrieving the robot should be straightforward. If it gets stuck or stops working, especially inside equipment, we should be able to get it out easily to avoid further damage.

8.1.3 Motion Scheme of Search and Rescue Robot

According to existing research, search and rescue robots have various drive and motion modes, including tracks, wheels, crawling, and wriggling etc. Table 8.1 shows the comparison and difference of these motion modes.

It can be seen from Table 8.1 that crawl mechanism is too bulky to enter the narrow space, making it unsuitable for the narrow space search and rescue. The wheel mechanism struggles with obstacle navigation, lack stability, and are easy to tipping. Additionally, their limited traction cannot support the transportation of essential supplies to survivors. On this basis, we take the wriggling mechanism to the design of slender rescue robot, its advantages include:

Table 8.1 Comparison of several different movement modes

	Drive	Advantage	Disadvantage
Tracks	Motor	Strong obstacle overcome ability, fast forward speed	Large size, can not enter the narrow space
Wheels	Motor	Simple structure, high reliability, moving fast on flat surface	Poor obstacle navigating ability, poor stability, easy to topple
Crawling	Pneumatic, motor, etc.	Small size, Strong obstacle overcome ability	Low traction, slow moving
Wriggling	Pneumatic, hydraulic, motor, etc.	Soft structure, good stability, strong passing ability, large traction	Complex structure design, and the slow moving

1. Stability
   Wheeled and legged robots can tip over easily in rough terrain due to instability. Peristaltic robots, with their low energy state, are less likely to fall. Plus, if designed with circular symmetry, they can keep moving even if they overturn, offering better stability.
2. Large traction
   Traction is the force a robot needs to move forward. For wheeled and legged robots, traction depends on their weight and friction. This creates a challenge: to increase traction, we need to increase the robot weight, which makes it harder to navigate obstacles. Conversely, the traction of a peristaltic robot isn't tied to its weight. Instead, it depends on its design, like its support and drive structures. This design advantage lets peristaltic robots remain lightweight while still offering good traction.
3. Compact structure
   Peristaltic robots are simpler than legged or wheeled robots, they don't need complex parts like wheels or tracks. They can use innovative materials like silicone rubber for movement and can be powered in different ways, like pneumatic or hydraulic. Their compact design makes them ideal for navigating narrow spaces in ruins or debris.

The inchworm is a type of invertebrate that moves by extending and contracting its body, similar to a peristaltic crawling motion. It stretches its front section forward and then brings its rear section up to meet it, as shown in Fig. 8.1. This movement is periodic, with the front and rear sections serving as stabilizers while the middle part propels the motion. The movement of the inchworm robot, as shown in Fig. 8.2, can be divided into the seven steps:

1. The rear retainer 2 opens until it supports the external environment.
2. The middle drive extends.
3. Rear retainer 1 opens until it supports the external environment.
4. Retract retainer 2 to its original state.
5. The middle drive is retracted to its original state.
6. Retainer 2 holds open until it supports the external environment.
7. Retract retainer 1 to its original state.

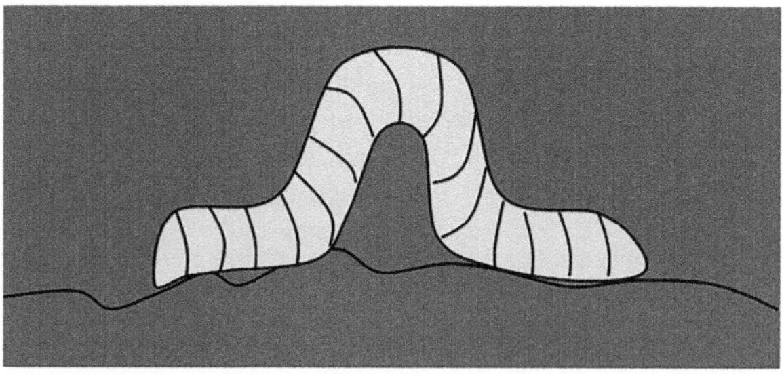

Fig. 8.1 Inchworm

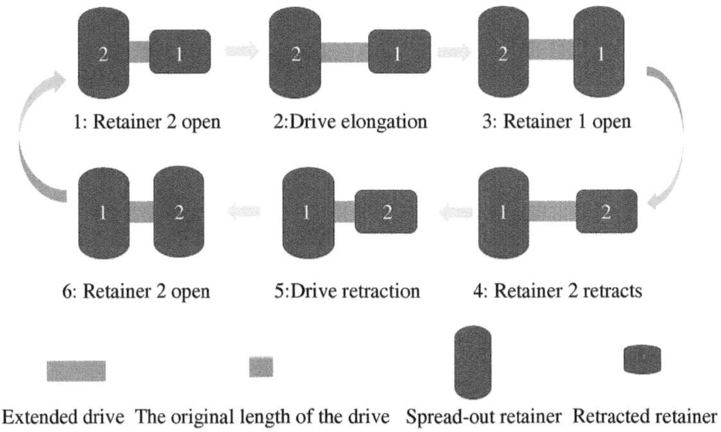

Fig. 8.2 Motion gait of the peristaltic crawling inchworm (Reprinted from [10] with permission from IEEE)

On this basis, we design a slender robot with the mentioned considerations. As Fig. 8.3 shows, the search and rescue robot includes the control platform, the drive mechanism, the crawling mechanism, the end manipulator and the feeding mechanism. The working process of the whole system is as follows: the search and rescue robot goes deep into the ruins by the feeding mechanism. In this process, the camera located at the robot end captures images of the internal environment, and transmitting these visuals to the PC for analysis. Using these images and sensor data from the robot, rescuers can locate survivors, assess their condition, and provide immediate help before additional rescue teams arrive.

The crawling mechanism is an insect-like crawling mechanism, which is not only used for movement guidance and traction, but also for posture maintenance and positioning. The flexible body acts like a hollow tube, allowing the robot to adapt to curved and intricate environments. The drive mechanism houses the actuator, controller, and power source. A tendon-sheath runs through the flexible body, enabling remote control of the end manipulator. Additionally, this tendon-sheath enhances the body stiffness. Finally, the feeding mechanism drives the robot to move forward.

8.2 Design of the Search and Rescue Robot

8.2.1 Design of Support Mechanism

In contrast to typical environments like pipelines, ruins present a highly complex and irregular environment. The channels within these ruins vary significantly in size, ranging from 60 to 260 mm. As a result, any support structure must be flexible enough

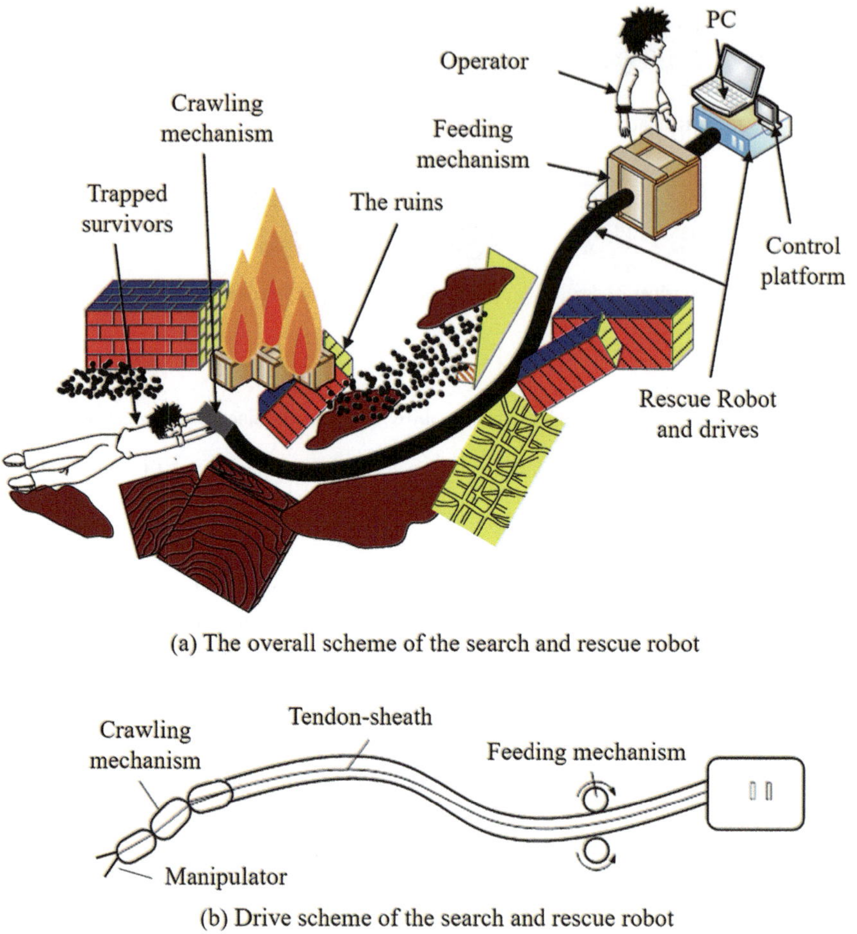

Fig. 8.3 Scheme of search and rescue robot system

to accommodate these irregularities and offer adequate support. To address this, two structural designs are proposed: the air bag structure and the scissor structure [9].

(1) **Airbag support structure**

Rubber and latex are flexible materials that work well for structures needing to fit different channel sizes. Similarly, airbags can adapt to uneven spaces in ruins. Figure 8.4 displays our designed airbag support structure, which consists of front and rear end covers, an airbag wall, and bellows. As gas fills a sealed space between the bellows and wall, the airbag expands while the bellows adjust, enabling a wide range of movement.

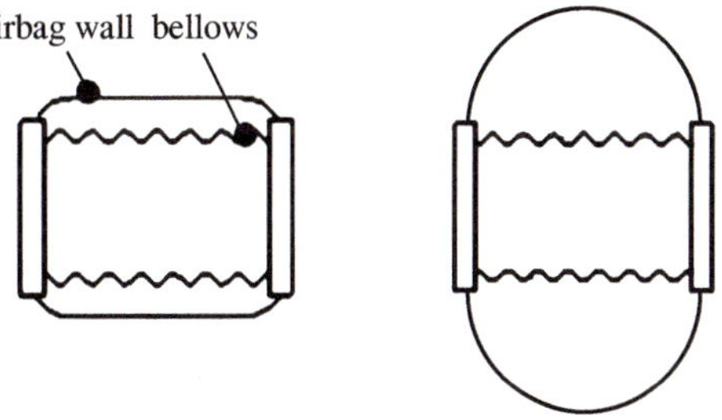

Fig. 8.4 Airbag support structure [9]

Fig. 8.5 Hydraulic lifting platform [9]

(2) Scissor support structure

The scissor structure is widely used in daily applications, like telescopic doors and hydraulic lifting platforms. As shown in Fig. 8.5, its biggest feature is that the telescopic size can be changed in a wide range such that the scissor structure can enable the search and rescue robot to adapt to the ruins channel with a wide range of size. Figure 8.6 displays our design of the scissor support structure, where six scissor mechanisms are evenly positioned on a cylindrical base.

Table 8.2 shows the comparison of the two support structures.

As shown in Table 8.2, the main advantage of the air bag support structure is its ability to adapt to various irregular channels in diverse environments. However, it has notable weaknesses. In the post-disaster environment, there are many sharp objects like steel bars or gravel that can easily puncture the airbag. To enhance the airbag design, one approach could be to use more durable industrial rubber or incorporate a steel wire mesh within the rubber layer. However, the production of industrial rubber is difficult and can not be completed in the laboratory. Additionally, adding a steel wire mesh complicates the process and may affect the flexibility of the airbag.

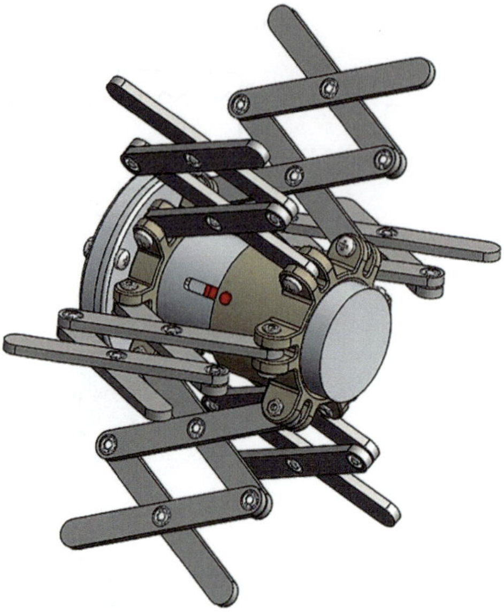

Fig. 8.6 Scissor support structure [9]

Table 8.2 Comparison of two support structures

	Airbag support structure	Scissor support structure
Driving mode	Pneumatic	Pneumatic, hydraulic, electrical
Ability to adapt to irregular channels	Strong	Weak
Elongation ratio	2.25	3
Supporting force	Smaller	Big
Deficiency	The airbag can tear easily on sharp objects	The round shape of the scissor structure doesn't fit well in uneven ruins passages

The scissor support structure offers a superior extension ratio compared to the airbag support structure and is more durable. Its main shortcoming is that the expanded envelope is circular, which limits its adaptability to irregular environments. Based on these considerations, we take an improved scissor support structure to construct the search and rescue robot.

As show in Fig. 8.7, we add small slider and driving spring into the original cylindrical slider of the scissor support structure. The small slider can move within the groove of the cylindrical slider. Both the small and cylindrical sliders are linked by a driving spring. As shown in Fig. 8.8, this modification results in an adaptive support structure where each scissor component can adjust independently with the

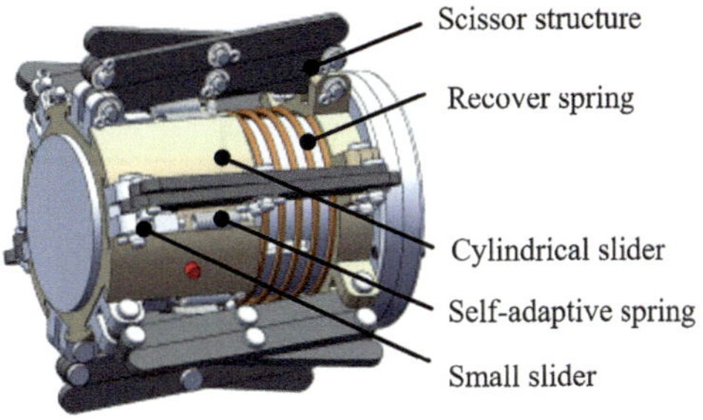

Fig. 8.7 Improvement of scissor support structure (Reprinted from [10] with permission from IEEE)

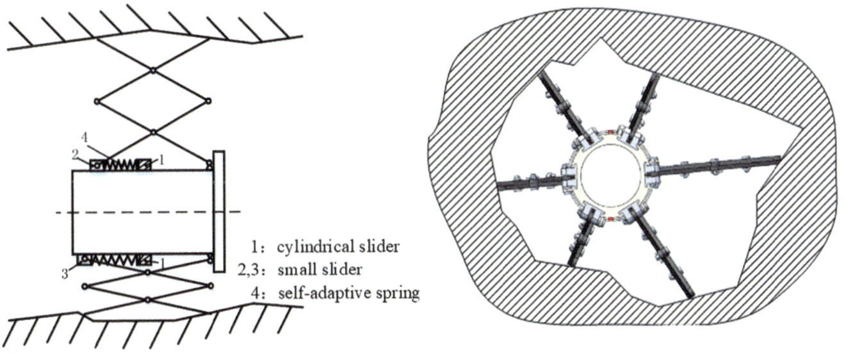

(a) Adaptive structure with spring

(b) Robot adapting to irregular passages

Fig. 8.8 Adaptive rendering (Reprinted from [10] with permission from IEEE)

Fig. 8.9 The sectional view of the supporting structure (Reprinted from [10] with permission from IEEE)

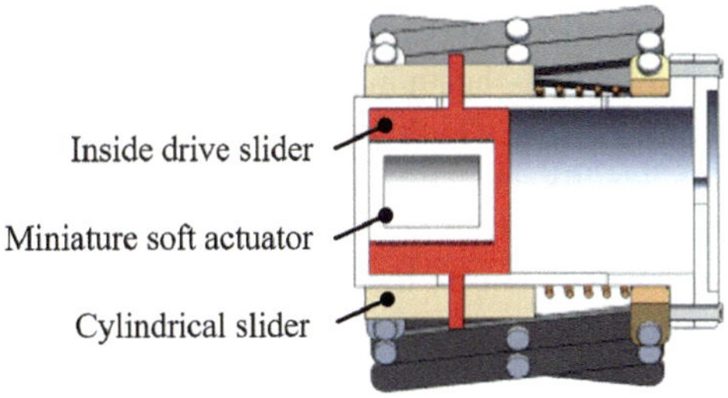

drive spring. This allows the structure to better navigate and adapt to irregular ruin environments.

The driving mechanism for the support structure is located inside, Fig. 8.9 is the sectional view of the supporting structure. The extension of the soft drive makes the slider to move forward. The cylindrical slider is fixedly connected with the drive slider, thus promoting the extension of the scissor structure. As a result, the scissor can expand from 80 to 240 mm.

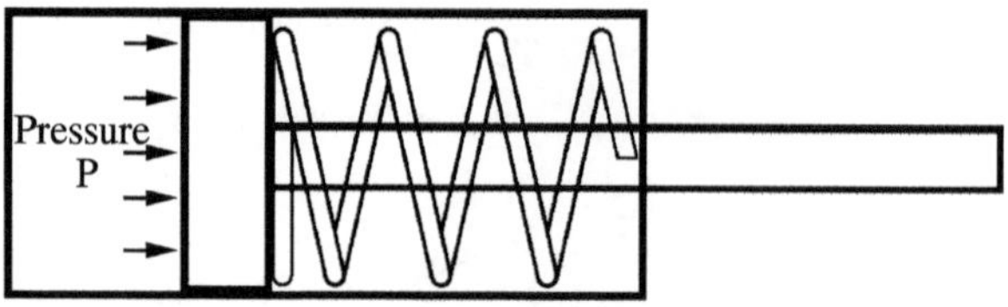

Fig. 8.10 Cylinder schematic [9]

8.2.2 *Design of Telescopic Structure*

To emulate the motion pattern of an inchworm, an intermediate driving structure for expansion is essential. Common scaling schemes are as follows.

(1) **Piston type cylinder structure**

The piston cylinder structure is a typical telescopic design, it propels forward by the pressure of the piston rod and restored to its original shape by a spring. As shown in Fig. 8.10, the cylinder, as a standardized pneumatic actuator, is simple and reliable. However, its rigid design struggles in navigating complex ruin channels, causing potential robot stuck. At the same time, the mass of the cylinder is larger, which will increase the burden of the robot entering the ruins.

(2) **Spring structure**

The series spring is flexible to adapt well to intricate channels. Figure 8.11 shows a series spring scheme designed for this project. The telescopic bending structure is formed by two springs in series and uses four tendon-sheath mechanisms for motion transmission. As shown in Fig. 8.12, the structure can flex and bend by controlling these tendon-sheath mechanisms. The tendon-sheath mechanism can realize remote transmission, which simplifies the structure design of the robot body and ensures the flexility. However, there are some difficulties in tendon-sheath transmission: the transmission friction is large and the tendon-sheath is easy to relax, which leads to inaccurate control. In addition, the self-weight of the long distance tendon-sheath is large.

(2) **Soft structure**

Research on soft structures has emerged in the past decade. These soft structures include silicone rubber, SMA (Shape Memory Alloys), and electroactive polymers, etc. These soft structures can passively alter their shape and size to suit various environments, such that they have safe and better environmental adaptability.

Figure 8.13 show the soft driver designed in this study. The body of the soft driver is made of silicone rubber, and the cylindrical channel in the middle is the trachea channel. The soft driver surface is evenly covered with Kevlar wires. Kevlar thread is a kind of fiber developed by DuPont in the 1960s. Compared with ordinary

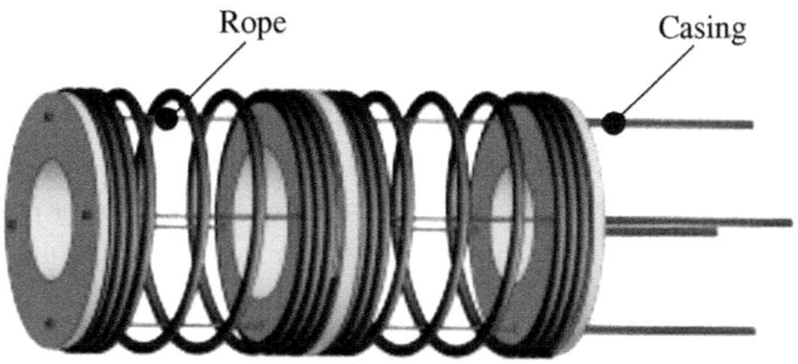

Fig. 8.11 Series spring structure [9]

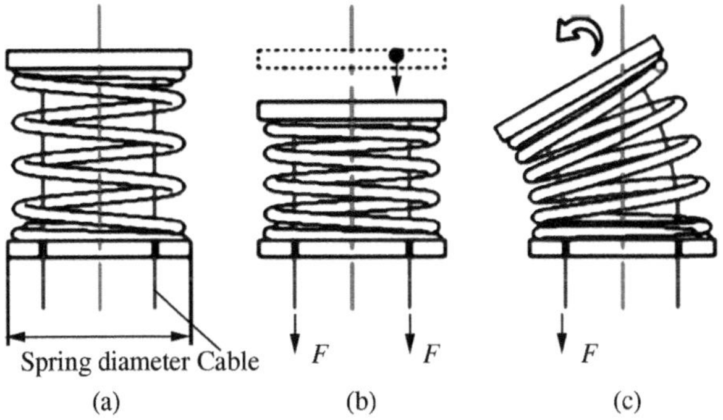

Fig. 8.12 Three states of series spring structure [9]

Fig. 8.13 Cross-sectional view of the soft actuator [9]

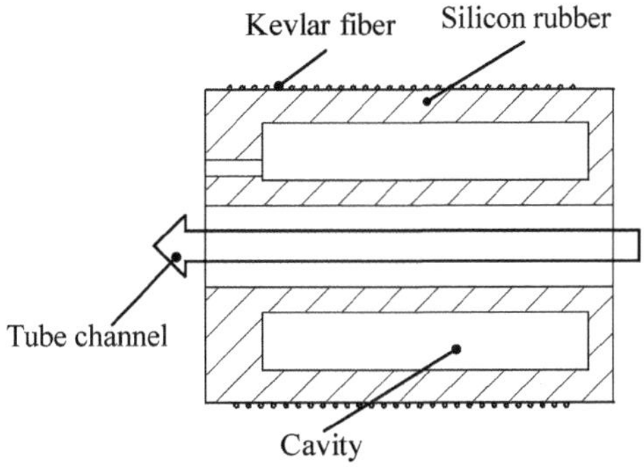

cotton thread material, it offers excellent flexibility, high tensile strength reaching 20.92 cN/dtex. For this project, a 0.38 mm diameter Kevlar wire was selected. This diameter effectively limits the deformation of the silicone rubber without compromising its integrity. When compressed air fills the cavity, the soft actuator expands. However, the Kevlar wires surrounding its surface restrict radial expansion, allowing only axial extension. The dimensions of the soft driver are 50 mm in length and 40 mm in diameter.

The soft telescopic structure excels in adaptability and cost-effectiveness compared to traditional piston rod designs. Crafted from materials like silicone rubber and Kevlar wire, it offers better maneuverability in complex environments without the need for costly precision machining or lubrication. With a maximum elongation of up to 60%, this design automatically restores its shape after pressure release, eliminating the requirement for external springs or mechanisms found in piston rod structures.

8.2.3 Design of Bending Joints

The passageway into the ruins is usually curved. In addition to making the robot adapt to the external environment passively, the ability to actively change the direction of travel is also important. Therefore, it is necessary to design a bending joint such that the search and rescue robot can actively change its movement direction.

Figure 8.14 shows the designed bending joint with two degrees of freedom, resembling a cylinder with four internal cavities. This joint is constructed from silicone rubber and encased with evenly distributed Kevlar wire. When air fills a cavity, the joint bends accordingly (Fig. 8.15), and upon air release, it returns to its original form, allowing for two-directional flexibility. This design allows flexibility without damaging the joint, offering a simple, reliable structure at a low cost.

8.2.4 Overall Structure of Soft Search and Rescue Robot

Figure 8.16 shows the overall design of the soft search and rescue robot, it mainly consists of three parts, the scissor support structure on both sides, the soft two-degree-of-freedom bending joint in the middle, and the soft driver [10]. The overall length of

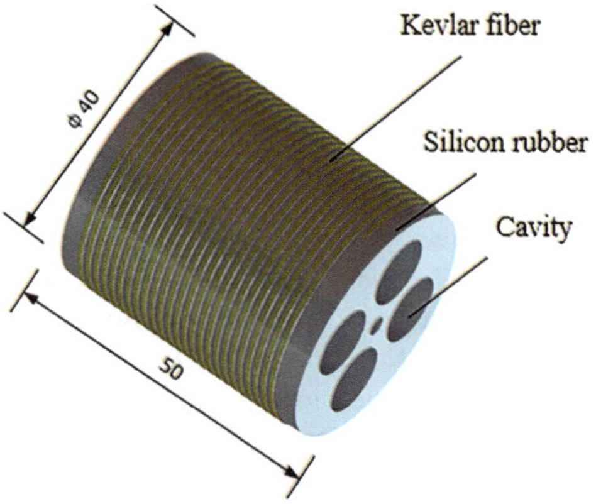

Fig. 8.14 Soft flexural joint [9]

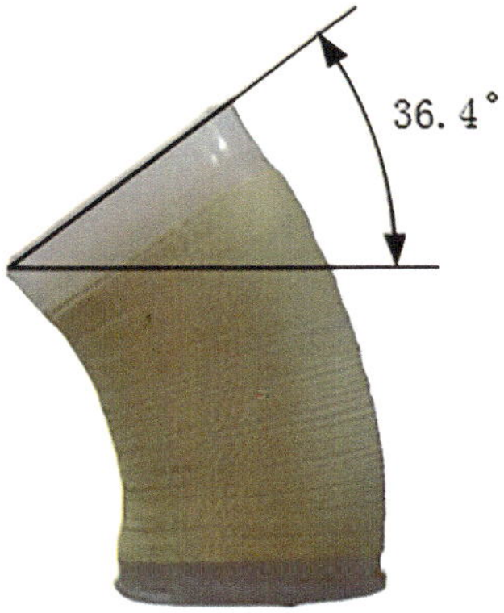

Fig. 8.15 Flexion state of the soft joint [9]

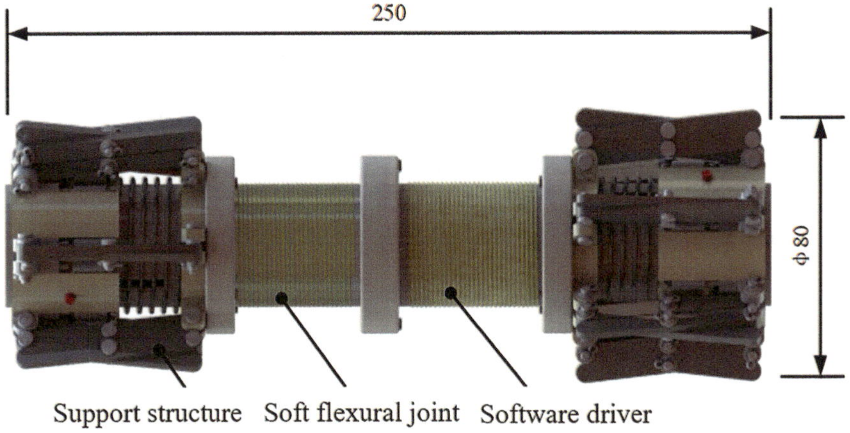

Fig. 8.16 Soft search and rescue robot [9]

the robot is 250 mm, and the minimum passable diameter is 80 mm. The soft search and rescue robot is driven by air pressure. To enhance its aesthetics and prevent air pipe interference with movement, the design incorporates internal air channels, as illustrated in Fig. 8.17.

8.3 Detection System with Inspection Experiments

To assess the performance of the rescue robot, an artificial disastrous environment was built with stones bricks and gravels. The passages built varied in width from 100 to 280 mm, with small stones serving as obstacles. Figure 8.18a–h show the peristaltic motion of the rescue robot in the experiment. It started with the expanded

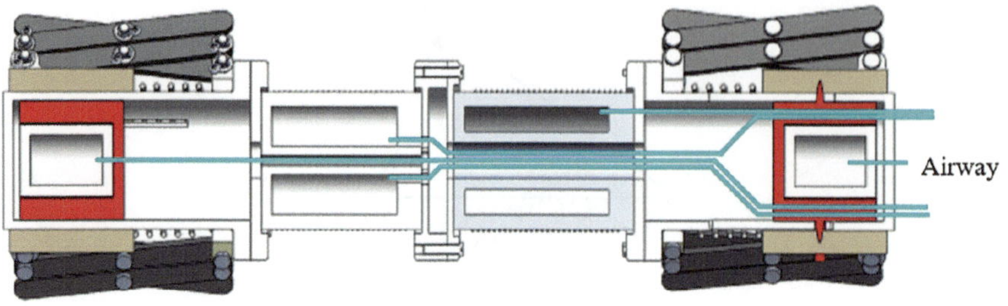

Fig. 8.17 Cross-sectional view of the soft robot [9]

posterior gripper (shown in Fig. 8.18a) and ended with the expanded posterior gripper after moved forward (shown in Fig. 8.18f), afterwards the rescue robot entered the next cycle (shown in Fig. 8.18g). Figure 8.18h captures the rescue robot navigating a curved passage with the angle of 130°. In general, the rescue robot can move forward in the passage at a speed of around 100 mm/min. As the passages in the ruins are complex and irregular, the rescue robots may encounter passages with uphill or downhill. To exam the mobility of the rescue robot in inclined passages, the climbing experiment is carried out. As Fig. 8.18i shows, the passage is a smooth one-meter-long acrylic tube with a diameter of 143 mm. Experimental results indicated the robot could climb inclines up to 25°, although the smooth surface of the tube posed challenges for the stability of the gripper.

8.4 Monocular Vision Positioning System

With the development of electronic technology and integrated circuit technology, the size of the vision sensor has been made to the millimetre level, but for slender robot applications, to build a binocular vision system that can effectively sense depth information, it is often necessary to increase its diameter, which affects the adaptability of the robot application. For this reason, slender robots often use a single camera to build their visual sensing system.

The single camera visual sensing system has a simple structure and is easy to configure and install, so it is more suitable for search and rescue environment robot application requirements. However, due to the limitation of the imaging principle, monocular vision cannot directly obtain depth information in principle. In order to obtain the depth information of the object, the range sensor can be used in combination with the monocular camera, which requires a fixed combination of range sensor and image sensor installation. In machine vision, use of a parallax for depth estimation is not novel [11]. Most studies have examined the parallax that emerges during large movements of a camera such as those produced by a vision system mounted on a mobile platform [12, 13]. These conditions amplify the motion parallax. However, the method of moving the camera and obtaining an image with parallax to calculate

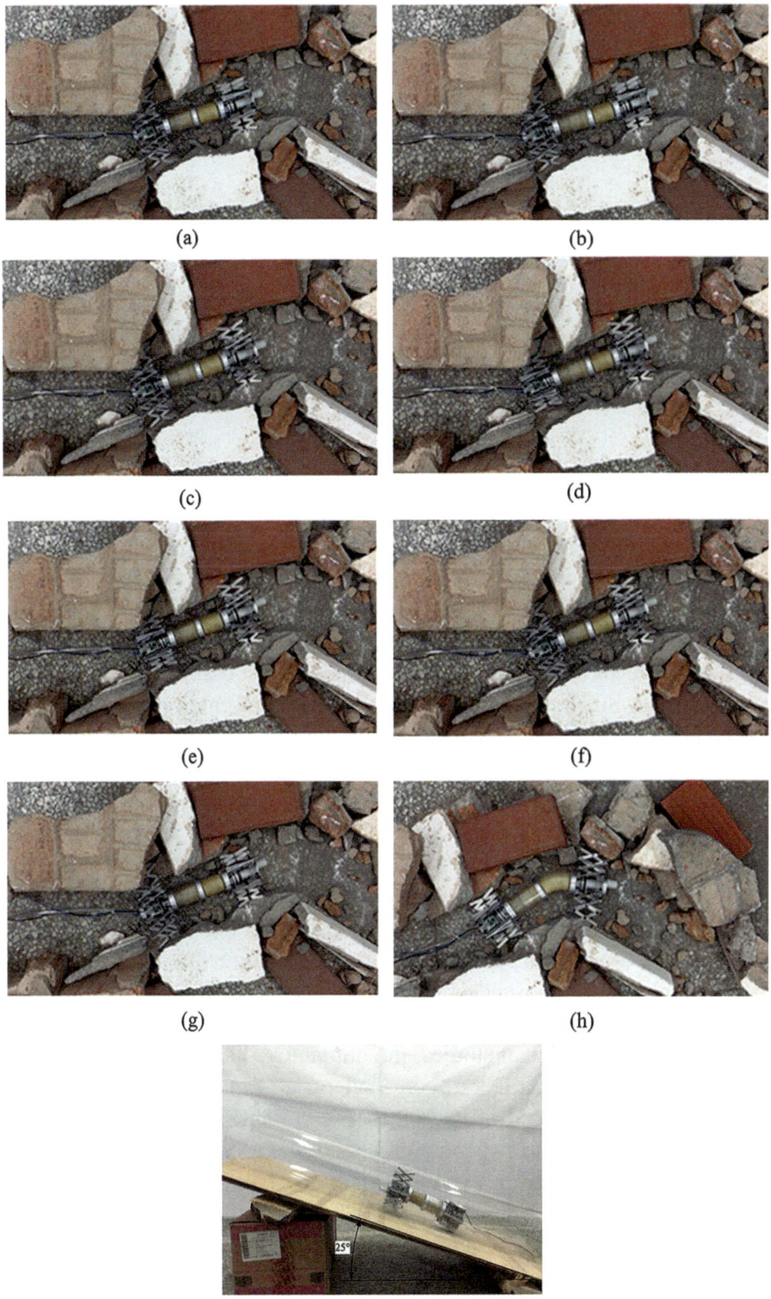

Fig. 8.18 Construction of a cable-driven rotating monocular vision system (Reprinted from [10] with permission from IEEE)

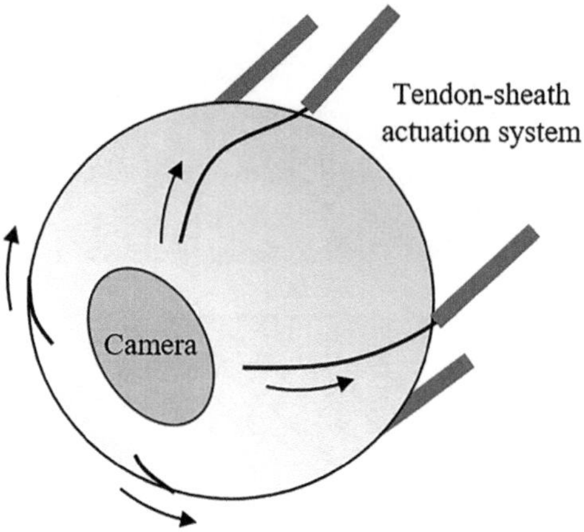

Fig. 8.19 Construction of a cable-driven rotating monocular vision system

the distance increases the robot size and the complexity of the control mechanics in applications of elongated robots with relatively small head sizes.

Similar to human beings, binocular animals are generally able to perceive the surrounding environment and the depth information of objects in the image through monocular vision after losing one eye. Biological experimental studies and cognitive analyses have shown that monocular animals can obtain depth information by collecting another image of the object through the combination of small positional gesture changes of the body and head, and small rotation of the eyeballs, i.e., to obtain two images in two positions with the same eye, and obtain depth information of the object similar to that of binocular vision through these two images. This is because the optical node in the eye is not consistent with the center of rotation, the eye usually moves slightly to adjust the projection of the object on the retina for gaze localization. The amount of this movement is quantitatively related to the amplitude of the eye rotation and the distance between the object and the observer. Based on this principle, we can use the parallax produced by rotating the camera of a stationary head/eye system to form a pseudo-binocular image to obtain the depth information, in which the nodal points of the camera do not lie on the axes of rotations.

According to physiological studies, the human eye has a sphere with 6 external eye muscles, each of which has a rectus scleral attachment point on the iris and is connected to the other end of its bottom track, of which only 4 rectus muscles play an important role in the sweeping process of the eye. Accordingly, a simplified model of rectus muscle-driven eye movement can be designed (Fig. 8.19), where the eyeball is designed as a homogeneous sphere of radius R with three rotational degrees of freedom around its center, and the external ocular muscles are replaced by tendon-sheath mechanisms that are connected to a tension actuation system. The tendon-sheath drive mechanisms including the head motion drive and the end robotic arm motion drive of the slender robot can be unified and realised by the tendon-sheath drive integration, which simplifies the design and control of the slender robot structure and drive.

Another way of monocularly estimating the depth information of an object is to estimate the depth information of an object through the combination of an image and its previous empirical knowledge of that similar environment, and in fact animals use their previous empirical information to estimate the distance of an object during binocular observation. Therefore, we can also combine the image with neural network learning technology, and obtain the depth information through neural network training and recognition estimation [14].

The deep neural network method for target object recognition and position estimation under monocular vision conditions for slender search and rescue robots can be carried out as follows: firstly, the appropriate network structure and algorithm are selected, and improved and simplified according to the characteristics of search and rescue detection applications to achieve the detection and tracking of target objects. This provides pixel coordinate information for the subsequent position estimation algorithm. Then, combining the known a priori information of the target object and the principle of computer vision, the PnP model is constructed to quickly and accurately obtain the transformation relationship between the camera coordinate system and the world coordinate system, and finally solve the position of the object.

References

1. Delmerico J, Mintchev S, Giusti A et al (2019) The current state and future outlook of rescue robotics. J Field Robot 36(7):1171–1191
2. Lyu M, Zhao Y, Huang C et al (2023) Unmanned aerial vehicles for search and rescue: a survey. Remote Sens 15(13):3266
3. Yamauchi B (2004) PackBot: a versatile platform for military robotics. In: Defense and security. International society for optics and photonics, pp 228–237
4. Nagatani K, Kiribayashi S, Nishimura T et al (2011) Redesign of rescue mobile robot Quince toward emergency response to the nuclear accidental Fukushima Daiichi Nuclear Power Station on March 2011. In: IEEE international symposium on safety, security and rescue robotics
5. Cachia A, Nazmul Huda M, Liu P (2019) Development and evaluation of a novel robotic system for search and rescue. In: Taros et al (2019) Towards autonomous robotic systems: 20th annual conference, London, UK, July 3C5, 2019, Proceedings, Part II 20. Springer International Publishing, pp 370–382
6. Alon W, Howard H, Choset et al (2005) Design and control of a mobile hyper-redundant urban search and rescue robot. Adv Robot 19(3):221–248
7. Cornell W, Austin B, Ben B et al (2012) Design and architecture of the unified modular snake robot. In: 2012 IEEE international conference on robotics and automation river centre, Saint Paul, Minnesota, USA, May 14–18, pp 4347–4354
8. Galloway K, Haynes G, Ilhan B et al (2010) X-RHex: a highly mobile hexapedal robot for sensorimotor tasks. Technical Report, University of Pennsylvania
9. Chen J (2016) The research of a soft rescue robot. Southeast University
10. Wang X, Zhang Q, Shen D et al (2019) A novel rescue robot: hybrid soft and rigid structures for narrow space searching. In: 2019 IEEE international conference on robotics and biomimetics (ROBIO). IEEE, pp 2207–2213
11. Sawhney (1994) 3D geometry from planar parallax. In: 1994 Proceedings of IEEE conference on computer vision and pattern recognition. IEEE, pp 929–934
12. Mansour M, Davidson P, Stepanov O et al (2019) Relative importance of binocular disparity and motion parallax for depth estimation: a computer vision approach[J]. Remote Sens 11(17):1990

13. Mertan A, Duff D, Unal G (2022) Single image depth estimation: an overview. Dig Signal Process 123:103441
14. Ming Y, Meng X, Fan C et al (2021) Deep learning for monocular depth estimation: a review. Neurocomputing 438:14–33
15. Chen J, Wang X (2015) Conception of a tendon-sheath and pneumatic system driven soft rescue robot. In: Robot intelligence technology and applications, vol 3. Springer International Publishing, pp 475–481
16. Tian F, Wang X (2008) The design of a tendon-sheath-driven robot. In: 15th international conference on mechatronics and machine vision in practice Auckland. New-Zealand Dec 2–4, pp 280–284

Chapter 9
Tendon-Sheath Transmitted NOTES Robots

Abstract Surgical procedures like endoscopy and hysteroscopy often require precise work in tight spaces within the human body. This includes tasks like observing, diagnosing, and cleaning. To handle these tasks, various surgical robots have been created. Besides the popular Da Vinci surgical system, there's a growing interest in creating robots that cause less harm, known as natural orifice surgical robots. These robots need to be flexible, compact, and reliable. A tendon-sheath mechanism is a good fit for these requirements due to its flexibility. In this chapter, we discuss two examples of tendon-sheath surgical robots, covering their design, testing, and practical uses.

9.1 State of the Art on Natural Orifice Translumenal Endoscopic Surgery Robots

Minimally invasive surgery, also known as interventional surgery, typically involves making one or more small incisions on the patient's body surface, or relying directly on the body's natural orifices, to introduce surgical instruments through these incisions or orifices into the body. The surgery is then performed under the guidance of the visual imaging system. Compared to traditional open surgery, minimally invasive surgery offers advantages such as reduced trauma, faster postoperative recovery, improved surgical quality, and lowered medical costs, significantly alleviating patients' surgical discomfort.

At present, the development of medical devices is forming a new research field, the minimally invasive surgical robot. These robots mainly focuses on neurosurgery, heart repair, gallbladder removal, artificial joint replacement, plastic surgery, urology, minimally invasive diagnosis and treatment and other fields. Over the past three decades [1, 2], the minimally invasive surgical robot technology has developed rapidly giving rise to multiple commercial products, as shown in Fig. 9.1. In 1994, the AESOP minimally invasive surgical robot developed by the American company Computer Motion became the world's first minimally invasive surgical robot to receive FDA (Food and Drug Administration) operational certification [3].

X. Wang and Q. Zhang, *Tendon-Sheath Transmitted Light and Slender Robots*,
https://doi.org/10.1007/978-981-96-5581-6_9

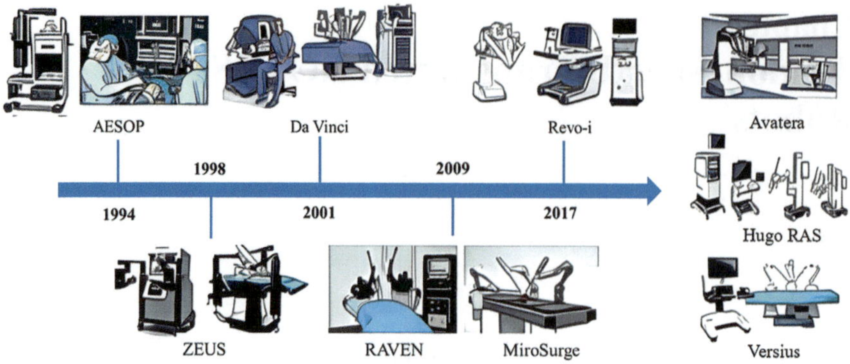

Fig. 9.1 History of minimally invasive surgical robots

In 1998, the company introduced the ZEUS surgical robot, which enabled master-slave operation in minimally invasive surgery [4]. In 2001, the American company Intuitive Surgical launched the renowned da Vinci surgical robot system, rapidly promoting its application [5]. The system is composed of the main console, robot car and serial surgical instruments. It adopts the master-slave control mode to control the robot movement. The main console, as the human-machine interaction interface between the doctor and the robot, is used to collect the doctor's hand movements and pedal movements to control the movement of the three mechanical arms and surgical instruments. The main console also provides doctors with three-dimensional images of the operation. By stepping on the foot pedal, doctors can switch the control of the robotic arm, adjust the posture of the master hand to an appropriate posture, and control the movement of the endoscope to align it with the lesion site and realize the zoom of the lesion area image. The surgical instruments installed at the end of the robotic arm adopt a wrist-like structure. Compared with traditional surgical instruments, this structure imitated human wrist greatly improves the flexibility of operation. The Da Vinci Surgical System represents the most advanced technology in minimally invasive cardiac surgery. As of now, the global installation of the da Vinci surgical system has exceeded thousands. Subsequently, minimally invasive surgical robots from around the world emerged successively, including RAVEN [6], Revo-I, Hugo RAS, and others.

With the development of minimally invasive surgical robots, it is the eternal pursuit of surgeons and patients to achieve the best treatment of diseases with minimal trauma (or even without trauma). In order to further achieve minimally invasive effects, some research institutions are also developing miniaturized robot systems with single channels [7, 8]. Compared with traditional minimally invasive surgery, single-channel minimally invasive surgery has less damage to the human body, and all surgical instruments enter the human body through the same surgical channel. One of the most typical examples is the transnatural endoscope surgical robot, or NOTES robot. Specifically, it refers to the type of surgery in which soft endoscope is used to enter abdominal cavity, chest cavity and other human body cavities through the channels of natural communication between human body and the outside world (such as mouth, esophagus, stomach, nodal (straight) intestine, vagina, bladder, etc.) for operation [9, 10]. For human body, NOTES mainly covers abdominal exploration, abdominal biopsy, liver examination, hysteroscopy, gastrointestinal and enteroenteric anastomosis, appendectomy, gallbladder resection, tubal ligation, partial hysterectomy, visual abortion, nephrectomy, splenectomy, pancreatic tail resection, etc.

An in vivo recombination endovascular surgical robot platform was developed by the Italian Institute of Bio-robotics Science [11]. Using in vivo recombination technology, the surgical robot platform consists of a fixed frame and three robotic arms, each of which has two degrees of freedom. In this way, the robot can be reassembled inside the body to perform surgery through the collaboration of multiple robotic arms.

A deformable robotic platform for intracavitary surgery has been developed at the University of Nebraska [12]. The intracellular surgical robot platform is composed of a robot body and two robotic arms, each of which has two degrees of freedom. The robotic arms are connected to both ends of the robot body through connecting rods, and the robot body is equipped with an endoscope and permanent magnet. The whole robot platform enters the abdominal cavity through the human esophagus in a straight line state. When it reaches the position, the main body of the robot is fixed on the abdominal cavity wall by external permanent magnet. The two robotic arms carry out surgical operation under the supervision of endoscope.

However, these types of single-port minimally invasive surgical robots typically employ rigid joints. While they provide strong operational force, they lack flexibility and cannot adequately meet the surgical requirements in confined spaces. Due to the separation achieved by cable transmission between the driving end and the operating end, the size of the actuators can be effectively reduced and the flexibility of the robot enhanced. Consequently, various cable-driven, elongated, multi-joint operating hands with continuous characteristics have emerged to meet the surgical demands of distant and multi-task operations within the body.

Purdue University developed a multifunctional endoscope robot based on tendon-sheath drive [13]. The entire surgical robot consists of an endoscope and two robotic arms, with a maximum external diameter of 13mm. Driven by a tendon-sheath, each arm has 9 degrees of freedom, making it extremely flexible and can be replaced with a variety of surgical instruments. The whole system has been made into a set of practical operating platform.

The use of tendon-sheath mechanism combined with continuum structures can achieve continuous bending and smaller sizes. Therefore, some NOTES robots use elastic single-skeleton continuums [14, 15], multi-skeleton continuums, soft robots [16, 17] to construct slender deformable actuators. In addition, there are also continuous structures such as pneumatic/hydraulic deformable joints [18] and multi-segment concentric tubes [19, 20] for minimally invasive robots. Compared to traditional multi-joint actuators, end-effectors based on continuous structures such as elastic skeletons and concentric tubes offer the advantage of smaller radial dimensions, along with flexible operation and good compliance. These are key research directions for future minimally invasive surgical robots, however, they also present more complex challenges in terms of sensing and control [21, 22].

Minimally invasive surgical robots can effectively improve surgical efficiency and reduce trauma, but the existing surgical robots have problems such as too large end instruments, soft operation and precise positioning, which make them unable to be used in neurosurgery, facial surgery, and pediatrics operate. In order to achieve less trauma and safe operation, precision, slender and compliant actuators with variable paths have been the constant pursuit of surgical robots.

9.2 Mechanical Consideration on Tendon-Sheath Transmission for NOTES Robots

Human abdominal cavity and the internal organs are a very complex environment, in the abdominal cavity distribution of various organs, each organ is closely connected, the abdominal wall is also tightly wrapped cavity organs. The focal site of the operation is also crowded with adjacent tissues and organs, and the operation process is easy to be interfered by other tissues and organs in the cavity, which on the one hand will reduce the flexibility of the surgical instruments, resulting in operation difficulties. On the other hand, it will hinder the endoscopic observation and affect the visual field of the operation. Therefore, it is difficult to operate on the lesions without causing surgical damage to any other organs and tissues of the human body by only relying on small surgical incisions.

There are also many problems when it comes to operating on the inside of human organs. The tissues in many organs of the human body are very elastic, normally tightly contracted together, and the inner walls are often not smooth, but wrinkled. Such as the human stomach, it is a solid pouch located in the upper abdomen below the lungs. On a natural empty stomach, the stomach of an adult is about the size of a fist, and the stomach walls form many irregular folds. Another example is the female uterus, it is located in the center of the pelvic cavity, between the bladder and rectum. In the absence of pregnancy, the lining of the uterus will fit together, also forming a large number of folds. If the lesions in the tissue occur in folds or are covered by folds, the endoscope entering the body will have difficulty finding these lesions, which will result in missed detection. And sometimes the organs constrict together, and there's very little room to maneuver, making it difficult to operate. Moreover, during the operation, the patient's cavity organs are under anesthesia, and the stimulation of the surgical instruments on the organ tissues during the operation often leads to the stress response of the organs, that is, the contraction or spasm of the organs. It is very dangerous for this kind of stress reaction to occur during the operation, which also causes obstacles to the smooth operation.

In order to solve the above difficulties in abdominal surgery, gas, liquid and air sac are used in clinical expansion of tissues and organs. For example, in abdominal surgery, the common practice is to establish an artificial pneumoperitoneum. Pneumoperitoneum is the key to the success of laparoscopic surgery. Artificial pneumoperitoneum should be established before the operation to separate the peritoneal wall from the viscera, expand the abdominal cavity to facilitate the operation, and avoid damage to the viscera during trocar puncture into the abdominal cavity. During hysteroscopic diagnosis and treatment, dilation technology is one of the most important and necessary key steps. During dilation, fluid medium such as CO_2 or saline is usually injected into the uterine cavity to dilate the uterus, so as to provide sufficient visual field and operating space for hysteroscopy.

However, there are many problems with using medium such as gas or liquid for expansion. Because gases and liquids are highly mobile, they are absorbed to some extent by the body's tissues. Studies have shown that the use of CO_2 gas for organ dilation is prone to complications such as acapnia, shoulder pain, nausea and vomiting [23], and the use of liquid for organ dilation is prone to electrolyte imbalance in patients [24]. When the airbag expands, it can become attached to the inner wall of the organ, preventing surgical instruments from performing live sampling and surgical excision. Therefore, the minimally invasive surgical robot needs a kind of organ dilator and a flexible manipulator, such that it can expand human organs without interfering with manipulator operation to assist its operation.

9.2.1 Design Methods of Tendon-Sheath Drive Expander

According to the characteristics of minimally invasive surgery, the following requirements should be met when designing the expander:

(1) The radial size of the expander in the closed state should be small enough to ensure that the expander can successfully pass through the incision of minimally invasive surgery or the natural cavity of the human body;
(2) The dilator should have sufficient stiffness after expansion, so as to prevent the surgical instruments from damaging the inner wall of the sub-organs due to the contraction of the inner wall of the organs during the operation;
(3) The dilator itself shall not harm human organs.

In view of the above design requirements, we introduce three design methods of the expander.

(1) **Vacuum rigid expander**

The expander in this scheme includes a plurality of spillage rods which are uniformly distributed in the circumferential direction and composed of rubber jackets and filled particles. In the closed state, the spillage rods have certain flexibility because the particles are free. During expansion, each rubber jacket is vacuumed, and the particles in the rubber tube are compressed together under atmospheric pressure, as shown in

Fig. 9.2. Depending on the friction between particles, the opening rod will obtain a certain stiffness to support the inner wall of the organ.

Figure 9.3 shows the working process of the solution. In the closed state, the expander is linear, as shown in Fig. 9.3a, with the expansion balloon in the middle and a number of extender rods evenly distributed around it. During the operation, the dilator is sent to the inside of the human organ, and compressed air is filled into the balloon through an inflatable tube connected to the dilator balloon. The dilator balloon will bulge into a ball, as shown in Fig. 9.3b, and the organ has been dilated. At this time, all the gas inside the opening rod is pumped out from the suction tube connected with the opening rod, so the opening rod is changed from a flexible rod to a rigid rod. When the gas in the dilating balloon is discharged, the balloon can be removed from the surgical instrument channel, as shown in Fig. 9.3c. At this time, the rigid extender rod assumes the task of organ dilation, and the surgical instrument can enter the trachea through the surgical instrument channel for operation.

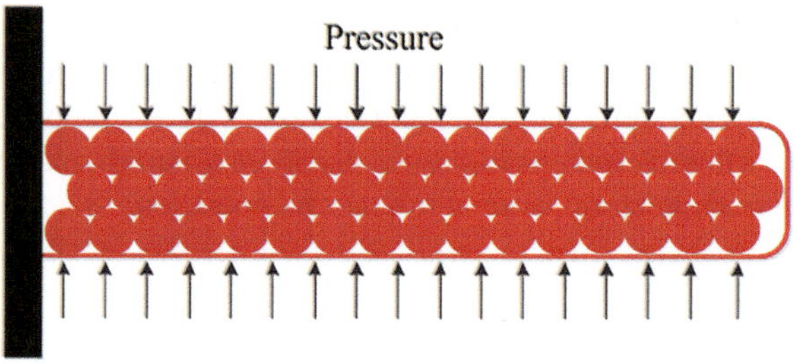

Fig. 9.2 Schematic diagram of the basic principle of vacuum hardening method [27]

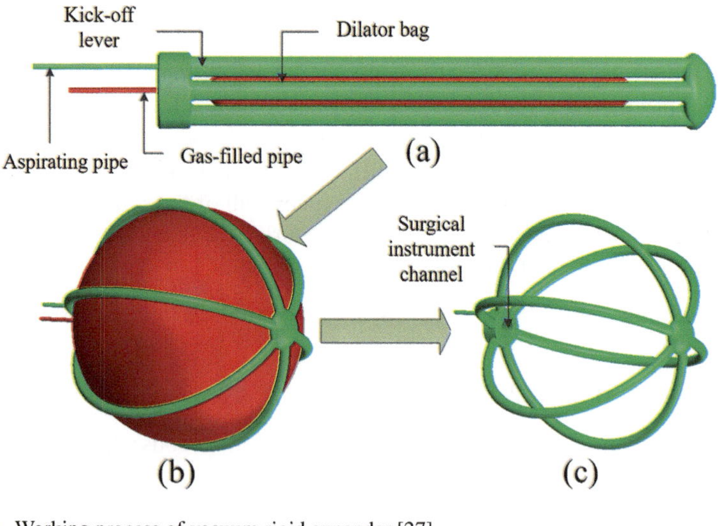

Fig. 9.3 Working process of vacuum rigid expander [27]

(2) **Flexible rod bending expander**

The expander in this scheme is mainly composed of a base, a connecting head, a plurality of flexible rods and drawpipes uniformly distributed in the circumference. The two ends of each flexible rod are respectively connected with the base and the connecting head through the flexible hinge. The drawing pipe is in the middle of the expander, and one end is connected with the connecting head, which can rotate around its axis. As shown in Fig. 9.4.

During the operation, the closed dilator is sent to the inside of the human organ, the base is fixed, the pull tube is pulled out, the flexible rod is bent into an arc shape, and the whole dilator forms a spherical shape, as shown in Fig. 9.5, so as to achieve the purpose of organ expansion. Due to the hollow structure of the tube-pulling tube, which is provided with tube-pulling slots and can be rotated 360°, endoscopes and surgical instruments can be inserted from the center of the tube-pulling tube and extended from the tube-pulling slots to observe and operate at various positions of the inner wall of the organ.

(3) **Multi-cell group expander**

In this scheme, each extender rod is composed of several acting blocks. The acting block is provided with a small hole, through which the tendon-sheath passes. Before entering the human body, the drive tendon of the expander in the closed state is in

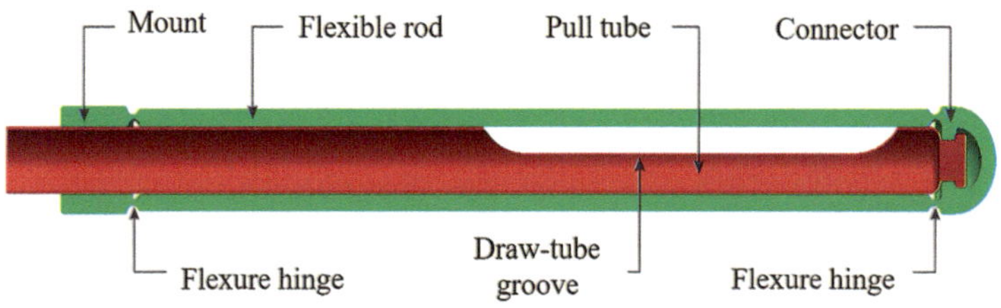

Fig. 9.4 Flexible bar expander (closed) [27]

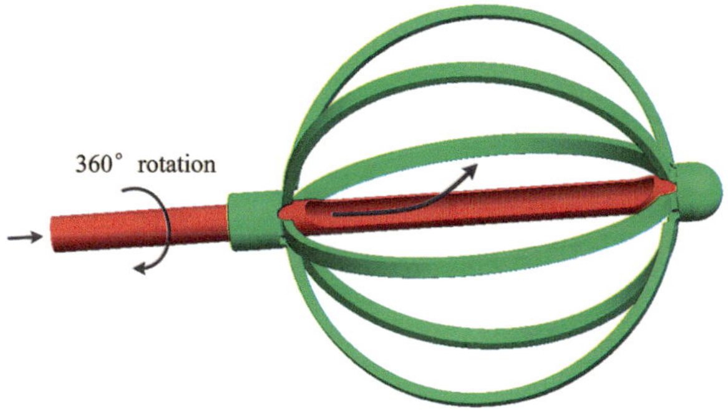

Fig. 9.5 Flexible rod spreader structure (after expansion) [27]

a relaxed state, and the extender rod has certain flexibility and is in a straight state. After the expander enters the organ, the tendon is pulled tight. The acting block is pressed together tightly under the action of the drive tendon, and the opening rod can obtain certain stiffness. The purpose of organ expansion can be achieved under the joint action of multiple opening rods. Figure 9.6 shows the basic principles.

The multi-unit group expander mainly consists of a fixed base, acting block, tendon-sheath mechanism, and expansion air bag. Figure 9.7 shows the specific working process. Before entering the inside of the organ, the dilator in the closed state is in a straight line as shown in Fig. 9.7. The dilator in this state is sent into the inside of the human organ, and the dilator air bag is inflated and expanded to stretch the organ. After tensioning all the tendons, each brace rod becomes rigid and assumes the task of organ expansion. The expansion balloon is removed from the surgical instrument channel, and the surgical instrument and endoscope can enter the internal organ through the surgical instrument channel for operation and observation. For the above three schemes, this section makes a comparative analysis of each scheme, and the analysis results are shown in Table 9.1.

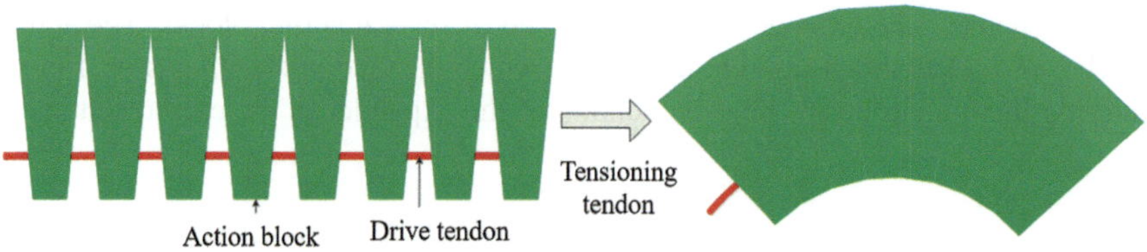

Fig. 9.6 Basic principle of multi-cell group expander [27]

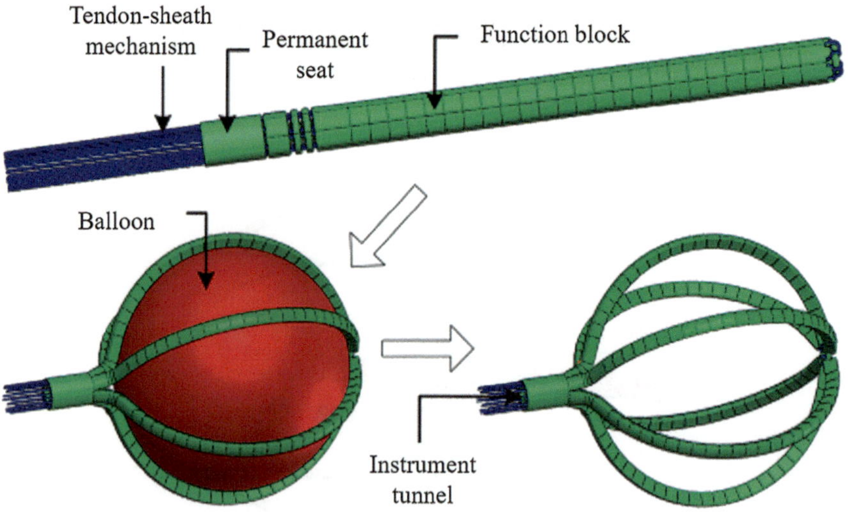

Fig. 9.7 Multi-cell group expander operation process [27]

Table 9.1 Scheme comparative analysis

	Advantage	Shortcoming
Scheme 1	The whole expander is composed of soft material, which has less damage to human body and high safety	The stiffness obtained by pressing the packed particles under atmospheric pressure is limited, so the opening rod needs to reach a certain diameter to obtain sufficient stiffness, which makes it difficult to reduce the radial size of the whole expander
Scheme 2	Flexible rod is used for organ expansion, with compact structure, small radial size and high stiffness	Due to the existence of pulling tube, surgical instruments can only be extended from the pulling tube slot for operation, which affects the operating flexibility of surgical instruments
Scheme 3	The spreader composed of multiple elements has both flexibility and rigidity. After expansion, the stiffness is large and the radial size of the whole expander is small	Each opening rod is composed of several units and depends on the tensioning tendon to obtain stiffness, which is relatively complicated in structure and operation

9.2.2 *Design Methods of Tendon-Sheath Actuated Operator*

More and more mechanical engineers have paid attention to the necessity of swinging the head at the end of the operator in the design of the natural human passage surgical robot. For example, when performing gastric cavity surgery, the most obvious advantage of the surgical robot compared with conventional endoscopic surgical instruments is that the robot can be designed with sufficient flexibility to satisfy the traversal observation, medical examination and treatment of the lesion location by the surgeon, which is conducive to improving the current disadvantage of the very low detection rate of gastric polyps by conventional gastroscopy. The front end of an endoscopy-carrying surgical robot can achieve two or three degrees of freedom, which is the most basic form to achieve such flexibility requirements.

In order to realize the swinging action of the operator end swing head, three types of swing operator mechanism are introduced here.

(1) **Two-way swing head**

As shown in Fig. 9.8, the bidirectional swing head is composed of multiple elements, and there are gaps between adjacent elements. Four tendon-sheath mechanisms are fixed on the lens holder. When the four tendons are pulled, the size of the gaps between the elements changes, and the endoscope at the end will swing at a certain angle. The angle at which the endoscope swings in both directions can be changed by controlling the displacement of each tendons.

The biggest advantage of this scheme is that it is flexible, but the device channel cannot be added to this structure under the limited size. The prototype produced according to this scheme is shown in Fig. 9.9, with a maximum diameter of only 6 mm.

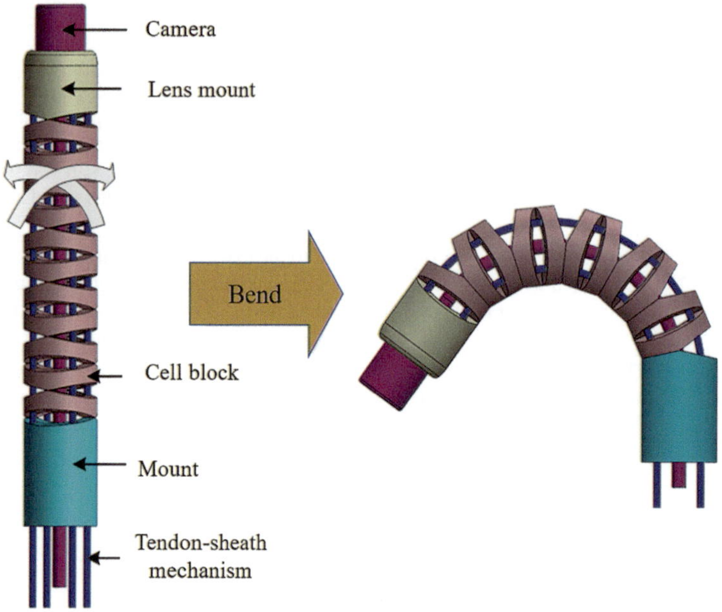

Fig. 9.8 Two way swing head [27]

Fig. 9.9 Bidirectional oscillating pendulum head prototype [27]

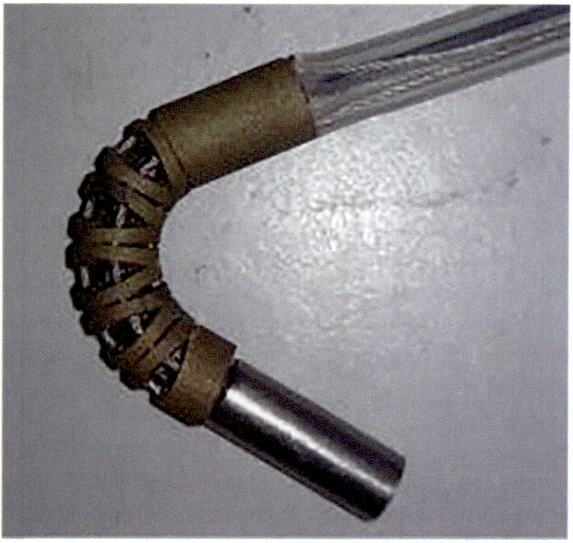

(2) **One-way swing type swing head**

As shown in Fig. 9.10, the structure of one-way swinging head is similar to that of two-way swinging head. Multiple units form the main body of the bendable part of the whole swinging head. The difference is that the structure can only bend in one direction in the plane. The pendulum head has two degrees of freedom: the degree of freedom to swing in one direction in the plane and the degree of freedom to rotate around the axis of the base. By tightening the tendon, the gap between adjacent units becomes smaller and the endoscope at the end of the swinging head can swing at an Angle. Release the tendon and the swing head will return to its original straight shape under the action of the reset spring.

The unidirectional swing head adds a surgical instrument channel, which shares a channel with the return spring. The swinging head can bend the biopsy forceps inside the instrument channel together. Because the biopsy forceps can also move along its axis, when the swinging head is in place and the end of the biopsy forceps is aligned with the lesion site, the biopsy forceps can be extended from the instrument channel and moved to the lesion site for operation.

Although the scheme is not as flexible as the bidirectional swing head, the surgical instrument channel is added under the same diameter limitation, through which

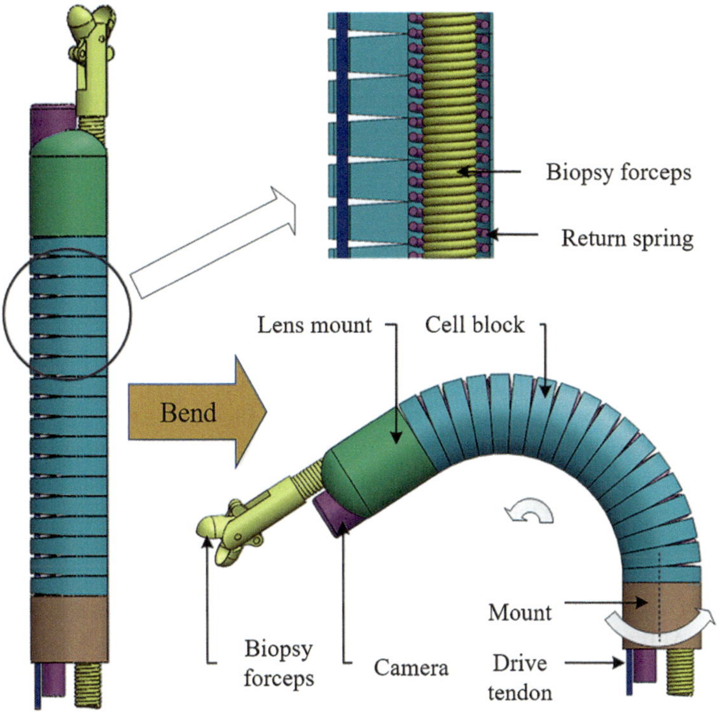

Fig. 9.10 One-way swing head

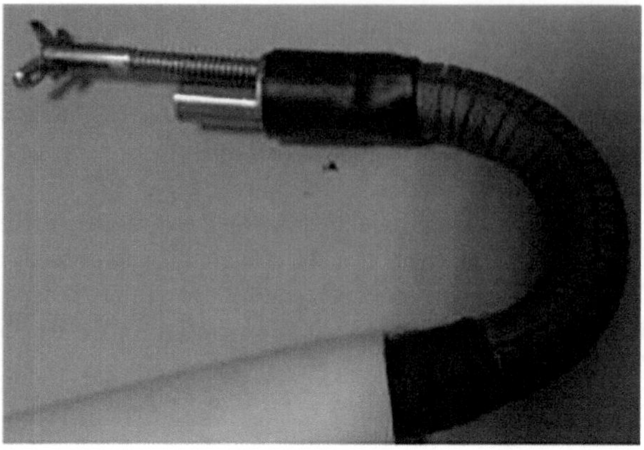

Fig. 9.11 Unidirectional oscillating head prototype [27]

surgical instruments such as biopsy forceps and negative pressure tube can enter the cavity for operation. Figure 9.11 shows the prototype manufactured according to this solution, with a maximum diameter of only 6 mm.

(3) **Restraint casing type swing head**

Figure 9.12 shows the operation principle of the restraint casing swing head. The pre-bent Ti-Ni alloy wire is located in a rigid restraint tube. The angle of the Ti-Ni alloy wire end can be changed by controlling the length of the Ti-Ni alloy wire extending out of the restraint tube. Figure 9.13 shows the structure of a restraint casing swing head. One end of the titanium nickel alloy wire is securely connected to the lens holder, and the other end is securely secured to the rightmost handle. The slider is fixedly connected with the restraint tube. However, when the slider is at the left end, the pre-bent part of the Ti-Ni alloy wire is completely in the restraint tube, and the Ti-Ni alloy wire is forcibly restrained into a straight line, as shown in Fig. 9.13a. Move the slider to the right, and the Ti-Ni wire protrudes from the restraint tube. Since the Ti-Ni wire is super elastic, the unrestrained Ti-Ni wire returns to its pre-bent shape, as shown in Fig. 9.13b, and the endoscope swings at an Angle. The swing angle of the endoscope can be controlled by controlling the displacement of the slider.

The oscillating principle of the surgical instrument is similar to that of the endoscope. The elastic tube is in a rigid sleeve, the elastic tube is prebent into a certain shape, and the middle of the elastic tube is the surgical instrument channel. When the prebent part of the elastic pipe is completely in the casing, the elastic pipe is forcibly constrained into a straight line, as shown in Fig. 9.13a. After extending the elastic pipe out of the casing, the elastic pipe recovers its pre-bent shape, as shown in (b) in Fig. 9.13. At this time, the end of the surgical instrument in the middle of the elastic tube swings at a certain Angle, and the surgical instrument can move along its own axis to the site of the lesion for surgical operation.

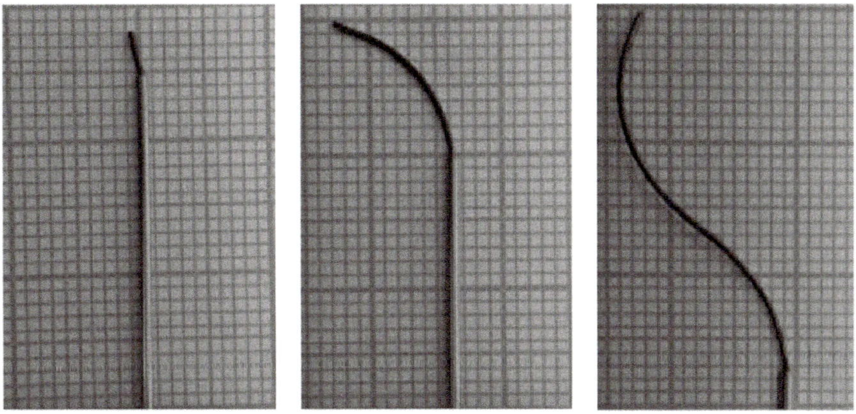

Fig. 9.12 Ti-Ni wire-casing construction [27]

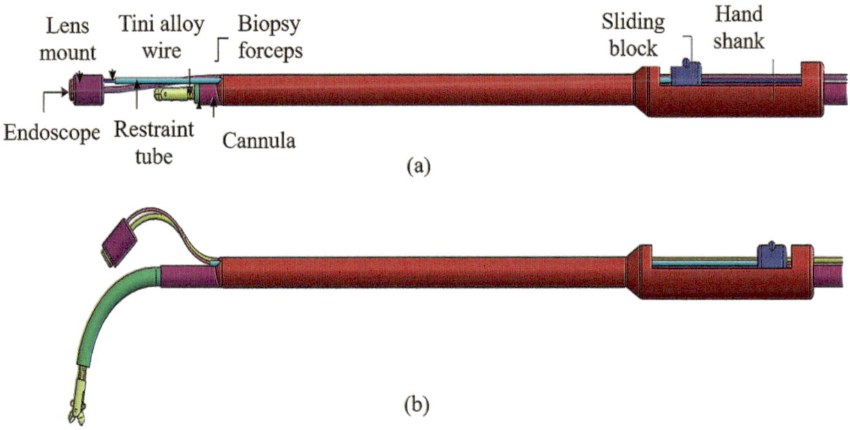

Fig. 9.13 Restraint casing swing head [27]

Fig. 9.14 Restraint casing oscillating head prototype [27]

Figure 9.14 shows the prototype produced according to this solution. The diameter of the operator is only 5.8 mm.

9.3 A Tendon-Sheath-Based Dilatation Device for Hysteroscopy

Hysteroscopy is regarded as the golden standard for the therapeutic and diagnostic methods of many uterine diseases. Carbon dioxide, normal saline and pharmaceuticals are generally used to dilate the uterus to obtain enough operating space and clear vision during the surgery. However, these methods often cause various syndromes. Compared with these methods, the mechanical methods, such as the balloon-based devices and the osmotic dilator have the lower risk of hyper stimulation.

However, in addition to the discomfort, poor integrity and tensile function, low flexibility and narrow operating space are other adverse characteristics in the application of balloon-based devices. To address this, we designed a novel mechanical dilator with a diameter of 9 mm at its initial status to pass through narrow cervix after initial cervical dilation by cervical dilator. Its diameter could be expanded up to 60 mm at the working status to achieve a favorable operating space. The operating system is composed of an endoscope and the surgical instrument driving tube. The endoscope was motioned by pre-bent hyperelastic wires and the surgical instrument was driven by a pre-bent driving tube. To obtain the parameters for successful expansion and operation, the relationships between the tension, the diameter of the dilator and the visual and operating space of the operating system were analyzed in detail [25].

9.3.1 Structure Design

The dilator was developed according to the structure of multi-cell group expander. The dilator unit is composed of many function blocks which has a hole on each bottom that allows a tendon-sheath to pass through it, such that all function blocks are connected together. When the tendon-sheath mechanism was relaxed, the dilator unit was flexible and longitudinally arranged in a line. Once the dilator was inserted inside the cervix, the rope was tightened and every function block was compressed together for the purpose of forming an arc arrangement. The unit would convert into a rigid body to maintain this shape. To expose much space as possible in one operation, the width of the single component of the rigid body was set to 3.2 mm. To avoiding shading the pathological area, two expansions were required. When the first expansion and operation were completed, the rope was relaxed. The dilator was rotated about 30° on the axis of the permanent seat. Then, the second expansion was performed, so that the hidden area would be exposed.

Figure 9.15 shows the three-dimensional structure diagram of the acting block. The section shape of the acting block is fan-shaped. There are three channels inside the acting block, namely two elastic wire channels and one rope channel. The elastic wire and the rope pass through several channels respectively, wherein the elastic wire can restore the straight shape after the rope is relaxed, so as to make the opening rod in a straight state.

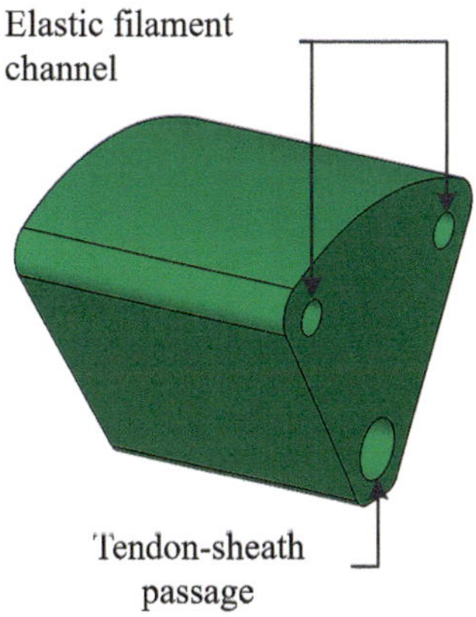

Fig. 9.15 Action block structure diagram [27]

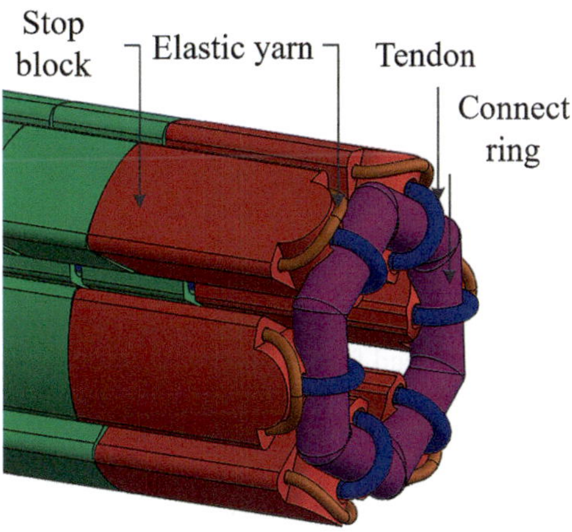

Fig. 9.16 Expander end connection structure [27]

Figure 9.16 shows the connecting structure of the end of the expander. Multiple opening rods are evenly distributed along the circumference, making the envelope surface of the expander cylindrical. In order to connect multiple stay rods together at the ends, a stop block and connection ring are added. In addition to three channels similar to the action block, a rope fixing channel is also provided inside the stop block to fix one end of the rope, and the other end of the rope passes through the rope channel by bypassing the connecting ring, connecting the stop block and the connecting ring together, and enabling the stop block to rotate around the connecting ring.

Figure 9.17 shows the root connection structure of the expander. In order to connect a plurality of open rod and fixed pipe, a connecting block, a transition block and an intermediate block are arranged at the connection, and each part is connected in turn. The structure of the transitional block and the intermediate block is similar to that of the acting block. The difference lies in the length of several parts and the angle between the two ends. An inclined plane is configured on the connecting block, as

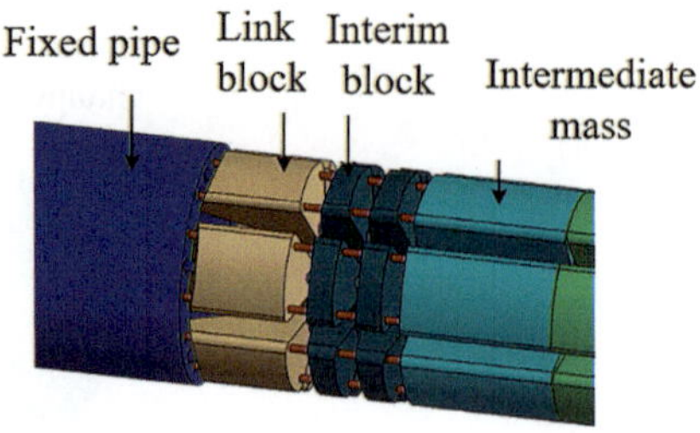

Fig. 9.17 Expander root connection structure [27]

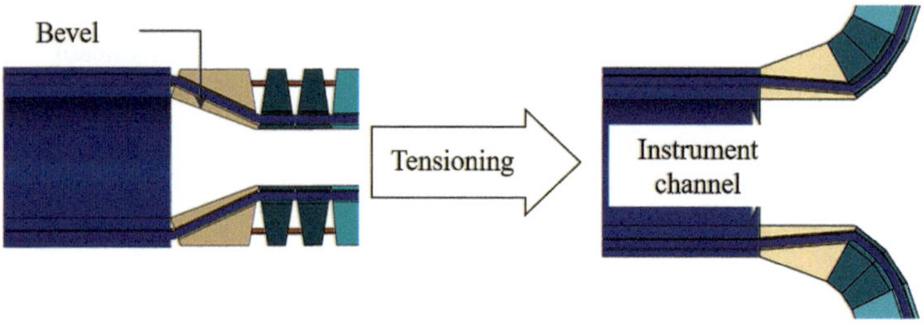

Fig. 9.18 Connecting block bevel [27]

shown in Fig. 9.18. After the dilator is expanded, one end of the connecting block is fitted to the end face of the fixed tube after rotation at a certain angle, leaving room for the central instrument channel.

Figure 9.19 shows the shunt structure of the instrument channel and the sleeve channel of the dilator. In this dilator, both the cannula and the instrument channel are axial. Multiple flexible cannula can block the instrument channel and prevent the instrument from entering the internal organ for surgical operation. Therefore, this section designs a channel shunt structure as shown in Fig. 9.19. The structure consists of an end cap, a base, a fixed block, and a shell. By taking advantage of the flexibility of the casing, the casing is bent 90° and then routed out of the casing channel, so that the instrument channel and the casing channel do not affect each other. However, because the casing is bent at a large Angle, it will inevitably increase the friction between the tendon and the casing.

The operating system was composed of two parts with a similar motion mechanism: the endoscope and the surgical instrument driving tube. The surgical instrument was located inside the driving tube. The hyperelastic wire (made of Nitinol) and driving tube was pre-bent into a certain shape (Fig. 9.20a). The hyperelastic wire was inserted into the constraint tube as Herrell et al. The angle and the location of the hyperelastic wire and the driving tube could be easily adjusted by changing their lengths outstretched the rigid tube or sleeve. The motion mechanism is demonstrated in Fig. 9.20b. One end of the hyperelastic wire was attached on the lens mount and the other end was attached on the hand shank. The whole hyperelastic wire was in the

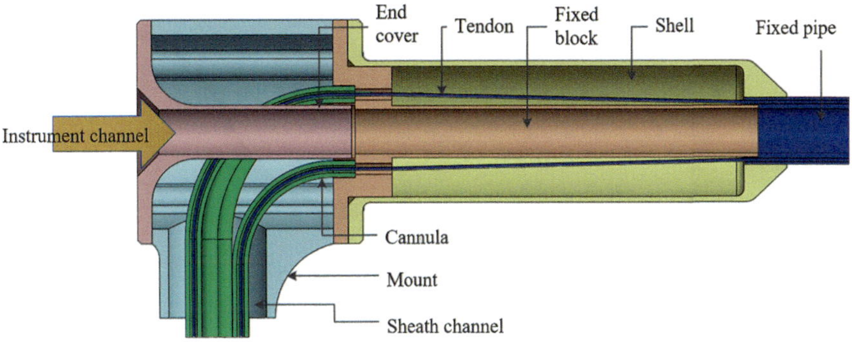

Fig. 9.19 Channel shunt structure of the expander [27]

rigid constraint tube and obligated into a linear shape as the sliding block was on the left. When the sliding block moved to the right, the hyperelastic wire was extended from the constraint tube and restored its pre-curved shape and the endoscope swung a certain angle. The device was achieved with the combination of the dilator and the operating system. The general view of this robot is shown in Fig. 9.20c.

The removing process of the dilator is demonstrated in Fig. 9.21a. The dilator could pass through the tunnel with a diameter of 10 mm easily. After the operation was completed, the ropes were relaxed to make the dilator become flexible. All the dilator units were movable and could restore their initial shape to produce the adaptive deformation, such that it could go through the narrow surgery channel, like the cervix.

The motion principle of the endoscope was similar to that of surgical instrument. In this experiment, the surgical instrument was biopsy forceps with a diameter of 1.75 mm. The biopsy forceps was placed into the pre-bent driving tube (with an inner diameter of 2.11 mm) and forced into a linear shape in the rigid sleeve. After extending from the sleeve, the driving tube could restore its original shape and the biopsy forceps could swing within a certain angle. The biopsy forceps presented different angles according to the length of the driving tube extending from the rigid sleeve. The movement process of the lens mount is shown in (Fig. 9.21b). The lens mount could swing different angles according to the length of the hyperelastic wire extended from the constraint tube. In order to realize a wider observation and operation scope, the whole operating system with a diameter of 5.8 mm could be rotated by rotating the hand shank.

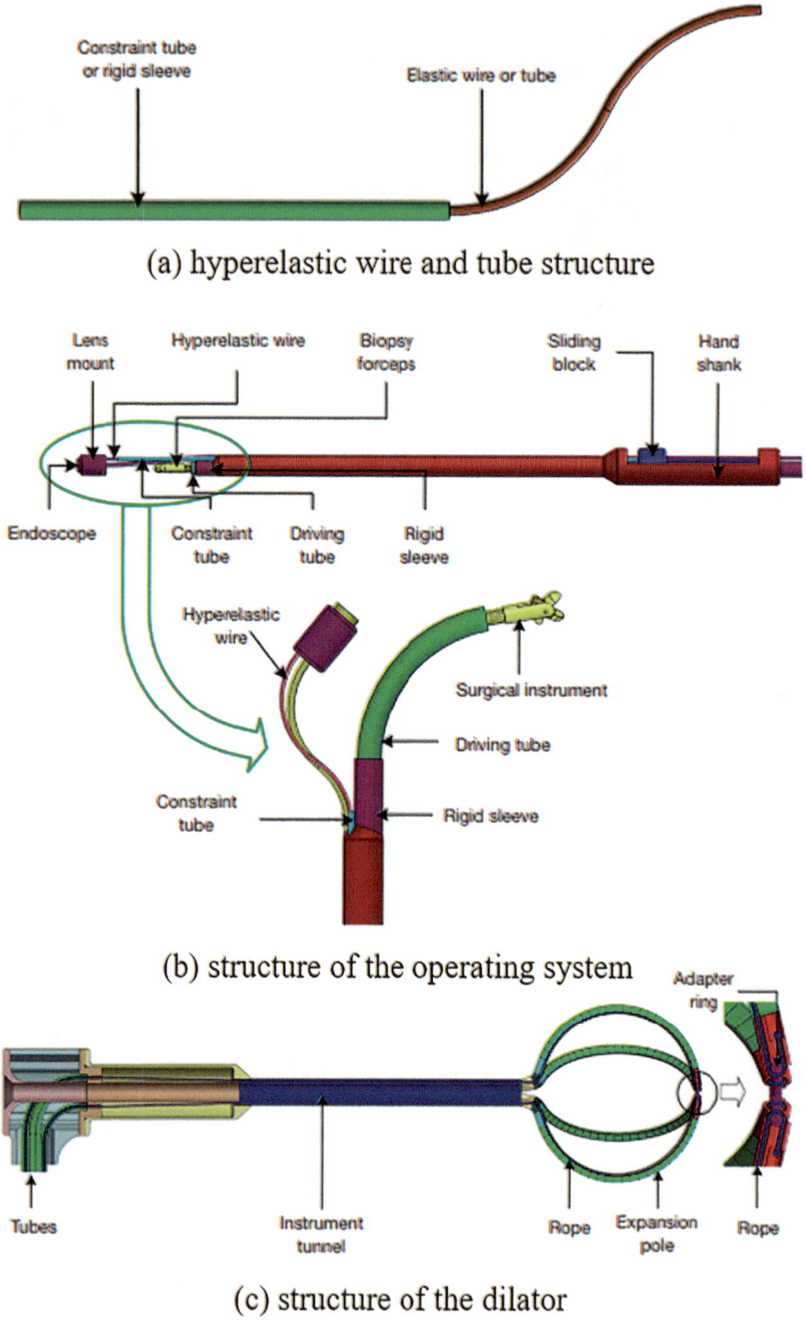

Fig. 9.20 The working principle and structure of the mechanical dilatation device (Reprinted from [25] with permission from AME)

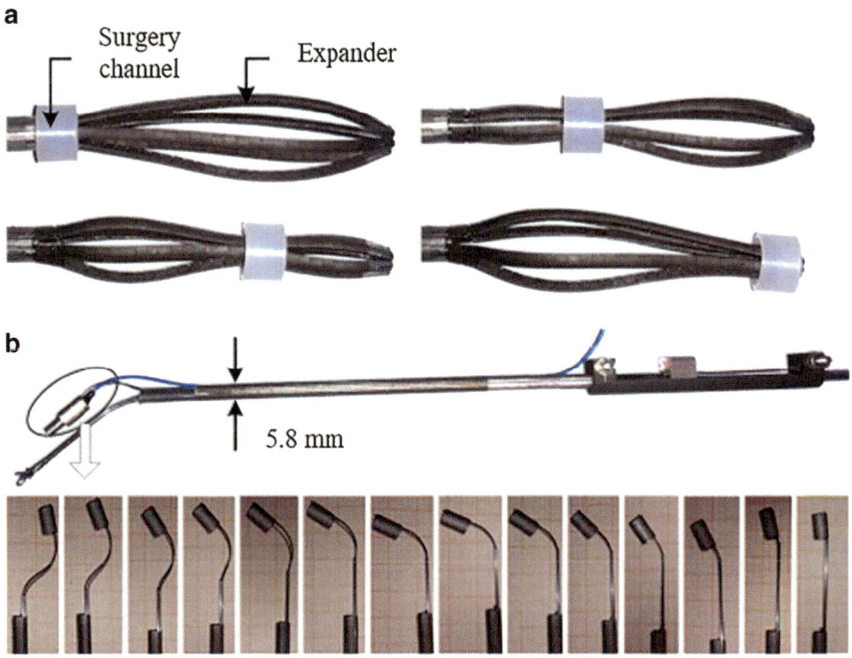

Fig. 9.21 The removing process of the dilator and operating system (Reprinted from [25] with permission from AME)

9.3.2 Simulated Hysteroscopic Micropolypectomy

In the simulated surgery, a raised small polyp was stick on the inner wall of the multilayer structure. The simulated uterus was expanded in three continuous steps with the dilator (Fig. 9.22a–c). After removing the balloon, the dilator acted like a cage to support the inner wall of the simulated organ and there was enough space in the cage, thus proving the functional feasibility of the expander. Then, the surgical instruments were thrust into the intracavity of the simulated uterus through the channel of the dilator. The raised polyp could be clearly observed and easily excised by the operator in such an adequate space. The process of the simulated surgery is shown in Fig. 9.22d–k. The whole process was repeated for 5 times and the average operation time was 18.37 ± 2.86 min.

The experiment verified the feasibility of the designed dilator and operating system in principle. However, there are also some limitations existed in this study. Due to the friction between the tendon and the functional blocks, the dilator units could not completely restore the linear state after relaxing the tendon. The structure design and processing technology of functional blocks should be further optimized in order to improve the dilator. Additionally, due to the huge difference in the uterine anatomic structure between human and large laboratory animals, we did not carry out an in vivo

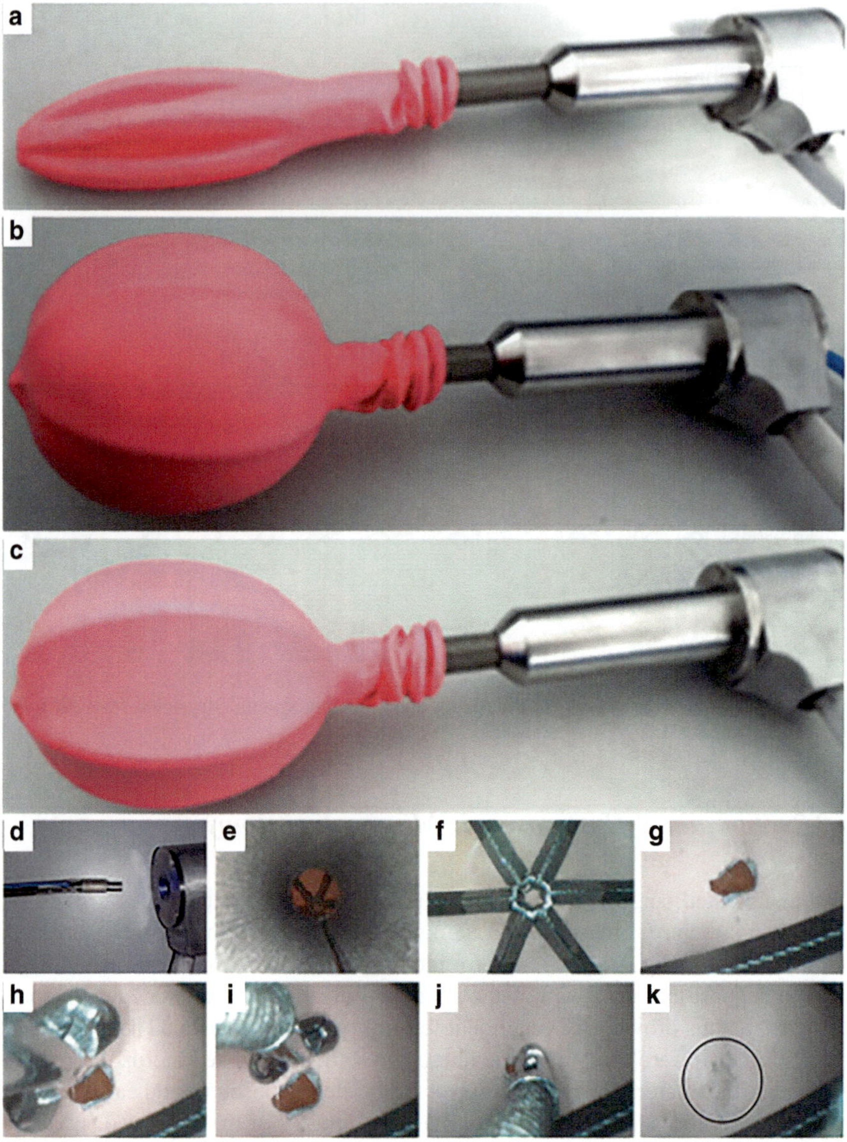

Fig. 9.22 Working procedure of the dilator and operating system (Reprinted from [25] with permission from AME)

experiment to validate the functionality and effectiveness of the dilator. Nevertheless, the study provides a new solution for the problems of organ expansion and operation in hysteroscopy.

9.3.3 Mechanical Analysis of the Expander

9.3.3.1 Deformation Analysis

Since human organs are soft structures, the interactions between the dilator and the inner wall of the organ as well as between the components of the dilator are very complex. Therefore, to simplify the analysis, the following assumptions are made in this section:

(1) When analyzing the deformation of the splitter, each small unit of the splitter is a rigid body;
(2) The shape of the bracing rod after deformation is either circular arc, and the arc length of the distribution of the applied force on the bracing rod is unchanged;
(3) The load of the inner wall of the organ acting on the opening rod is uniform load, and the stress of each opening rod is the same.

All of the following force analysis is based on these three assumptions. Suppose the initial tension of the rope is T_0 after the organ is stretched apart by the air bag, and the tension of the tendon becomes T after the air bag is removed.

Since it is complicated to calculate the elongation of the rope by using formula (9.3.1), it is simplified here. The rope is regarded as a linear spring. According to the mechanics of materials, the deformation of the rope under stress can be obtained:

$$\varepsilon = \frac{(T - T_0)\,l}{EA} \tag{9.3.1}$$

where, E is the elastic modulus of the rope material, A is the cross-sectional area of the rope, and l is the length of the rope.

When the extender rod is stretched out without pressure from the inner wall of the organ, the extender rod is simplified into the model as shown in Fig. 9.23, where R_0 is the initial outer arc radius, s_0 and l_0 are the initial outer and inner arc lengths respectively, β_0 are the central angles corresponding to the initial arc lengths, and h is the unit height.

When the opening rod is deformed by the load of the inner wall of the organ, its shape is still circular. Its radius, inside and outside arc length and the corresponding central angle to R, s, l, β respectively (Fig. 9.24).

Since all acting blocks are regarded as rigid bodies, the outer arc length s and element height h do not change before and after deformation. The inside arc elongation ε, namely $l = l_0 + \varepsilon$. Depending on the geometric relationship are: $R\beta = R_0\beta_0$,

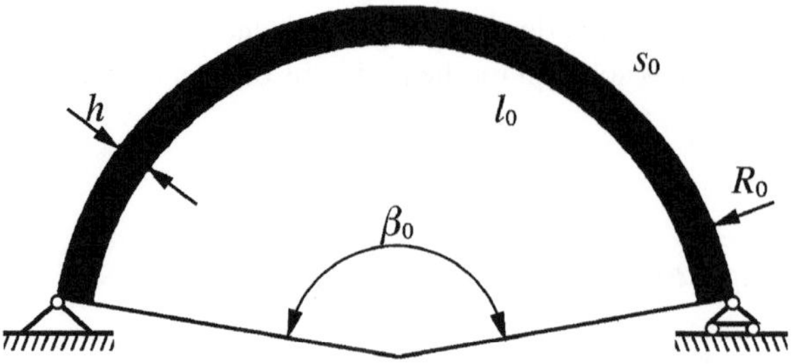

Fig. 9.23 The original shape of the opening lever [27]

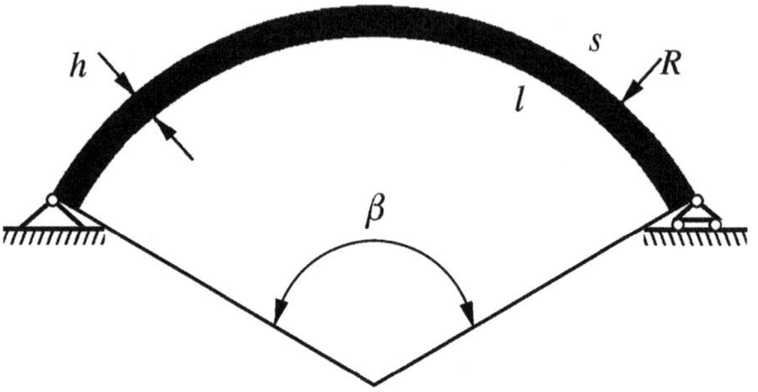

Fig. 9.24 The shape of the brace rod after deformation [27]

$\beta(R-h)=\beta_0(R_0-h)+\varepsilon$, $R_0\varphi_0=R\varphi$, reduction available deformation formula is:

$$\begin{cases} R=\frac{hR_0\beta_0}{h\beta_0-\varepsilon} \\ \beta=\beta_0-\frac{\varepsilon}{h} \\ \varphi=\frac{R_0}{R}\varphi_0 \end{cases} \tag{9.3.2}$$

9.3.3.2 Force Analysis

The applied force of the inner wall of the organ on each opening rod is the uniform load f_0, and the direction of the uniform load is the normal direction of the stressed surface, as shown in Fig. 9.25. In the figure, $f_{friction}$ is the component of f_0 along the tangential direction of the extender rod. This component is offset by the friction between the organ and the extender rod and will not affect the shape of the extender rod. Only the component force f will cause the deformation of the extender rod. Due to different stress location component of f is not the same, in order to simplify the analysis, it is approximately considered that f is a uniformly distributed load, and its value is equal to f_0. Its distribution range is the arc length range corresponding to the central angle φ, therefore the stress is simplified as shown in Fig. 9.25.

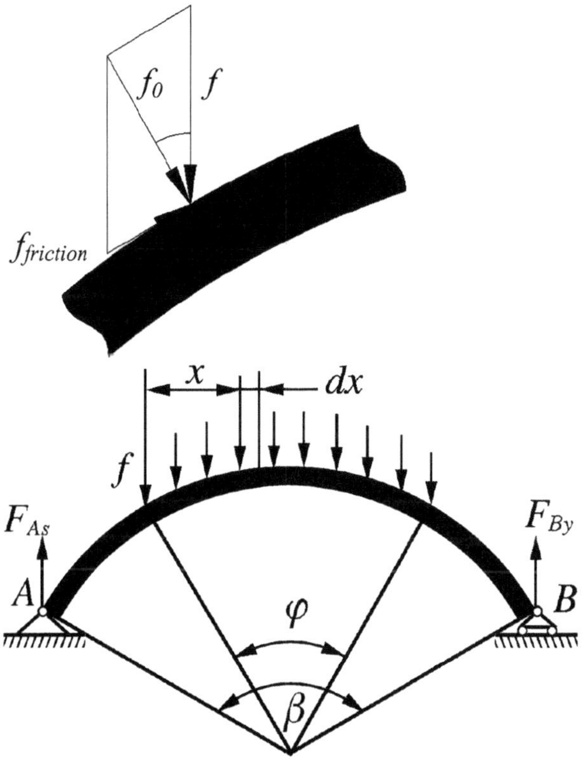

Fig. 9.25 Force analysis diagram of the opening rod [27]

Torque of the distributed load on fulcrum A:

$$M_A(F) = \int_0^{2R\sin\frac{\varphi}{2}} \left(R\sin\frac{\beta}{2} - R\sin\frac{\varphi}{2} + x\right) f dx \tag{9.3.3}$$

Component of uniform load in y direction:

$$F_y = 2Rf\sin\frac{\varphi}{2} \tag{9.3.4}$$

According to the balance equation of force and moment, it can be obtained:

$$\begin{cases} \sum F_x = F_{Ax} = 0 \\ \sum F_y = F_{Ay} + F_{By} - F_y = 0 \\ \sum M_A = 2F_{By}R\sin\frac{\beta}{2} - M_A(F) = 0 \end{cases} \tag{9.3.5}$$

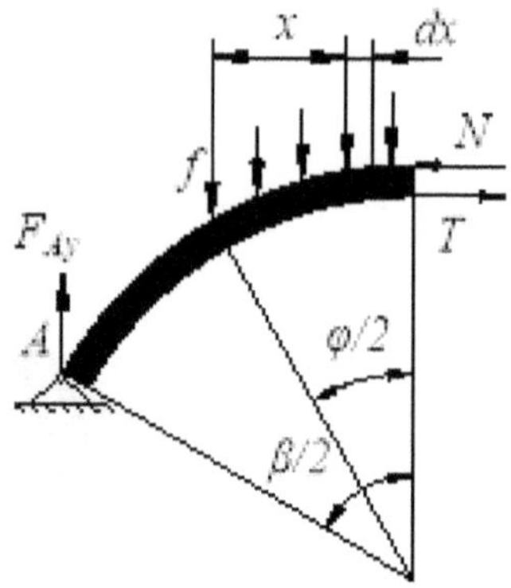

Fig. 9.26 Force analysis of 1/2 bar [27]

By simplifying formula (9.3.6), the bearing reaction can be obtained:

$$F_{Ay} = F_{By} = 2Rf \sin \frac{R_0 \varphi_0}{2R} \tag{9.3.6}$$

In order to analyze the internal force of the opening rod, the left half of the opening rod is taken for analysis. The pressure of the acting block on the left half of the opening rod is N and the tension of the tendon is T. The force analysis is shown in Fig. 9.26.

According to the balance equation of force and moment, it can be obtained:

$$\begin{cases} \sum F_x = T - N = 0 \\ \sum M_A = M_A\left(\frac{F}{2}\right) + T\left(R - R\cos\frac{\beta}{2} - h\right) - N\left(R - R\cos\frac{\beta}{2}\right) = 0 \end{cases} \tag{9.3.7}$$

where: $M_A\left(\frac{F}{2}\right) = \int_0^{R\sin\frac{\varphi}{2}} \left(R\sin\frac{\beta}{2} - R\sin\frac{\varphi}{2} + x\right) f dx$

The rope tension can be obtained as follows:

$$T = \frac{fR^2}{h} \sin\frac{\varphi}{2}\left(\sin\frac{\beta}{2} - \frac{1}{2}\sin\frac{\varphi}{2}\right) \tag{9.3.8}$$

By substituting formula (9.3.2) into formula (9.3.8), we get:

$$T = \frac{fhR_0^2\beta_0^2}{\left[h\beta_0 - \frac{(T-T_0)l}{EA}\right]^2} \sin\frac{R_0\varphi_0}{2R}\left[\sin\frac{EAh\beta_0 - (T-T_0)l}{2EAh} - \frac{1}{2}\sin\frac{R_0\varphi_0}{2R}\right] \tag{9.3.9}$$

According to the above equation, when the initial tension of the rope is T_0, the tension T of the rope under load can be obtained, and the radius R after

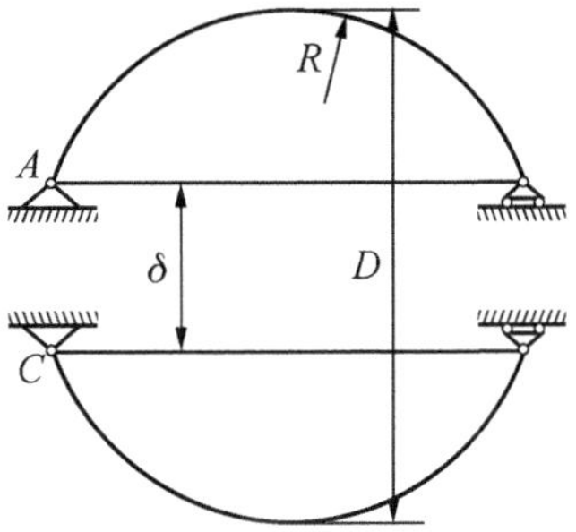

Fig. 9.27 Dilator diameter solution diagram [27]

deformation can be obtained. According to Fig. 9.27, the diameter of the expander can be calculated as follows:

$$D = 2R\left(1 - \cos\frac{\beta}{2}\right) + \delta \tag{9.3.10}$$

δ is the distance between the symmetric fulcrum, which is a fixed value.

9.3.3.3 Mechanical Analysis of Organs

Human organ tissues are soft and elastic, such that they can stretch when the organ expands. Due to the irregular shape and nonlinear properties of these tissues, the interaction force between tissues and dilators is hard to analyze. To simplify the analysis, the following simplifications are made in this section:

(1) The organ is reduced to an ideal sphere;
(2) The tension on the organ's inner wall is consistent whether it's stretched to its maximum shape or expanded to the same volume using liquid or gas;
(3) The organ is an incompressible linear material, whose shape can be changed arbitrarily, but the volume of the organ material remains unchanged.

A spherical crown with a height of h_0 was taken from the standard spherical organ wall. The radius of the circle with the maximum opening of the spherical crown was r_0, and the thickness of the organ wall was t_0. After aeration and deformation, the height of the spherical crown becomes h, the radius of the circle with the largest opening of the spherical crown becomes r, and the thickness of the organ wall becomes t, as shown in Fig. 9.28.

According to the geometric relation:

$$\begin{cases} h_0 = R_0(1 - \cos\theta) \\ h = R(1 - \cos\theta) \end{cases} \tag{9.3.11}$$

$$\frac{r_0}{r} = \frac{R_0}{R} \tag{9.3.12}$$

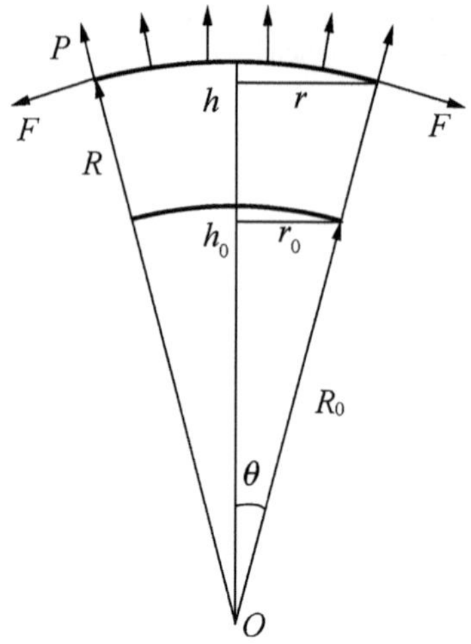

Fig. 9.28 Before and after organ wall deformation [27]

The formula (9.3.11) and formula (9.3.12) generation goal crown formula $S = 2\pi Rh$:

$$\frac{S}{S_0} = \frac{2\pi Rh}{2\pi R_0 h_0} = \frac{R^2}{R_0^2} \tag{9.3.13}$$

According to the hypothesis, the volume of organ tissue before and after deformation is unchanged, that is, $V_{tissue} = S_0 t_0 = St$.

Then the thickness of the organ wall after deformation is:

$$t = \frac{S_0 t_0}{S} = \frac{R_0^2}{R^2} t_0 \tag{9.3.14}$$

The cross-sectional areas of spherical crown organs before and after deformation were:

$$\begin{cases} A_0 = 2\pi r_0 t_0 \\ A = 2\pi rt = 2\pi \frac{R}{R_0} r_0 \frac{R_0^2}{R^2} t_0 = 2\pi r_0 t_0 \frac{R_0}{R} \end{cases} \tag{9.3.15}$$

The internal tension of the organ wall after deformation obtained from the mechanics of materials is:

$$F = \frac{A}{A_0} (A_0 - A) E_1 \tag{9.3.16}$$

where E_1 is the elastic modulus of the organ wall.

From the force balance equation:

$$F \sin\theta - P\pi r^2 = 0 \tag{9.3.17}$$

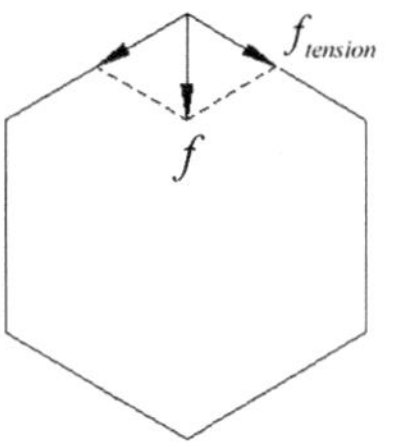

Fig. 9.29 Simplified relationship between the force on the opening rod and the tension on the inner wall of the organ [27]

In summary, the relationship between internal organ pressure and organ radius can be obtained:

$$P = 2E_1 t_0 R_0^2 \left(\frac{1}{R^3} - \frac{R_0}{R^4} \right) \tag{9.3.18}$$

The tension per unit length of the inner wall of an organ:

$$f_{\text{tension}} = \frac{F}{2\pi r} = E_1 t_0 R_0^2 \left(\frac{1}{R_0^2} - \frac{R_0}{R^3} \right) \tag{9.3.19}$$

The relationship between tension and pressure per unit length of the inner wall of an organ:

$$f_{\text{tension}} = \frac{PR}{2} \tag{9.3.20}$$

It is assumed that each rod has the same force. In order to simplify calculation, the tension f_{tension} of the rubber film and the tension f of the rod are simplified as shown in Fig. 9.29.

According to the simplified force, it can be obtained:

$$f = 2 f_{\text{tension}} \cos \frac{(n-2)\pi}{2n} \tag{9.3.21}$$

where n is the number of opening rods.

9.3.4 Kinematics Analysis of the End Operator

The designed operator comprises two main components: an elastic tube with position control and an oscillating endoscope. The elastic tube acts as a manipulator with three degrees of freedom. For effective operation, this manipulator must have the capability to reach any point on the an inner wall of organs. Robot kinematics is an important basis for the design or evaluation of robot motion performance. There are two main kinematics analysis method: forward kinematics determines the position of the tool end based on its design and movement, while inverse kinematics identifies how the components of the tool should move to reach a desired location. In this section, an effective mathematical method for describing robot pose will be provided based on the knowledge of linear algebra, engineering matrix theory and geometry.

9.3.4.1 The Coordinate System of the End Operator

There are many methods to describe manipulator coordinate system, including matrix method and vector method, among which matrix method is represented by D-H method. D-H method is a method proposed by Denavit and Hartenberg in 1995 to uniquely describe links and the position relationship between adjacent links with four parameters. The following is the general process of establishing manipulator connecting rod coordinate system according to D-H method.

First, in order to determine the relative motion and position relationship between the connecting rod, a coordinate system is fixed on each connecting rod. Where the base coordinate system 0 is fixed at the center of the top of the operating lever, and the z0 axis is collinear with the sleeve axis. The casing has the degree of freedom to move axially, so the casing coordinate system 1 is fixed at the center point of the top of the casing, and the z1 axis is collinear with the axis of the casing. Elastic tubes have two degrees of freedom: axial movement and axial rotation. To facilitate the analysis, two coordinate systems are established for the elastic tube. The initial position of the coordinate system 2 of the elastic tube coincides with the coordinate system 1. The coordinate system 3 is established at the center of the end of the elastic tube, and the z3 axis is collinear with the tangent direction of the end. The manipulator with established coordinate system is shown in Fig. 9.30.

9.3.4.2 Forward Kinematics Analysis

The linkage transformation, relative to the homogeneous transformation ${}^{i-1}_{i}T$ of the linkage coordinate system $\{i\}$ with respect to $\{i-1\}$, can be decomposed into the problem of four basic sub-transformations of the coordinate system $\{i\}$, each of which depends on only one link parameter. The four sub-transformations are:

$$
{}^{0}_{1}T = \begin{bmatrix} 1\,0\,0\,0 \\ 0\,1\,0\,0 \\ 0\,0\,1\,z_1 \\ 0\,0\,0\,1 \end{bmatrix} \tag{9.3.22}
$$

$$
{}^{1}_{2}T = \begin{bmatrix} \cos\theta_2 & -\sin\theta_2 & 0\,0 \\ \sin\theta_2 & \cos\theta_2 & 0\,0 \\ 0 & 0 & 1\,0 \\ 0 & 0 & 0\,1 \end{bmatrix} \tag{9.3.23}
$$

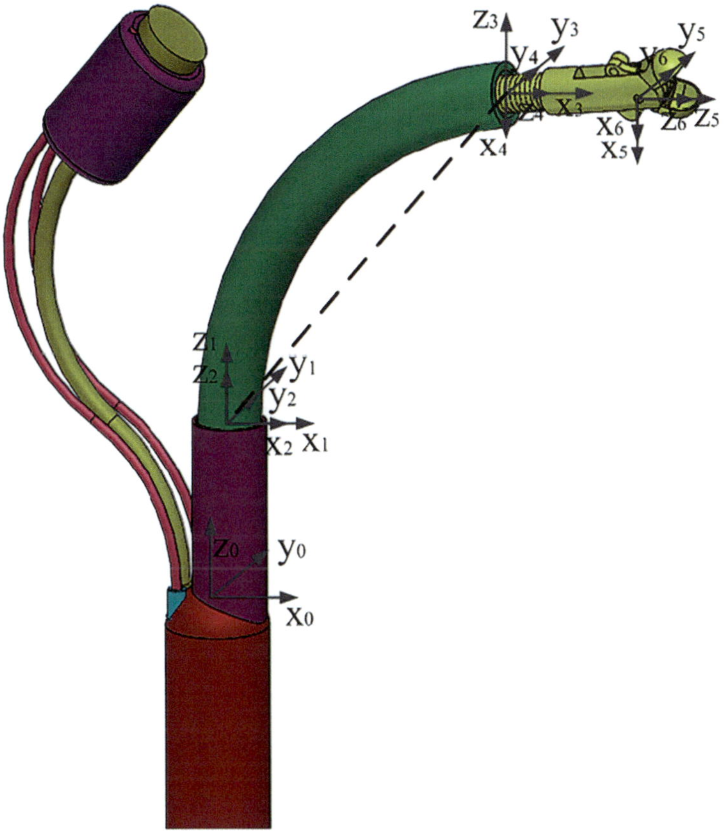

Fig. 9.30 The establishment of manipulator connecting rod coordinate system [27]

$$
{}_{3}^{2}T = \begin{bmatrix} 1 & 0 & 0 & R(1-\cos\theta_3) \\ 0 & 1 & 0 & 0 \\ 0 & 0 & 1 & R\sin\theta_3 \\ 0 & 0 & 0 & 1 \end{bmatrix} \tag{9.3.24}
$$

$$
{}_{4}^{3}T = \begin{bmatrix} \cos\theta_3 & 0 & \sin\theta_3 & 0 \\ 0 & 1 & 0 & 0 \\ -\sin\theta_3 & 0 & \cos\theta_3 & 0 \\ 0 & 0 & 0 & 1 \end{bmatrix} \tag{9.3.25}
$$

$$
{}_{5}^{4}T = \begin{bmatrix} 1 & 0 & 0 & 0 \\ 0 & 1 & 0 & 0 \\ 0 & 0 & 1 & z_5 \\ 0 & 0 & 0 & 1 \end{bmatrix} \tag{9.3.26}
$$

$$ {}^{5}_{6}T = \begin{bmatrix} \cos\theta_6 & -\sin\theta_6 & 0 & 0 \\ \sin\theta_6 & \cos\theta_6 & 0 & 0 \\ 0 & 0 & 1 & 0 \\ 0 & 0 & 0 & 1 \end{bmatrix} \tag{9.3.27} $$

where z_1 is the distance that coordinate system {1} moves along axis z_0, θ_2 is the rotation Angle of coordinate system {2} about axis z_1, θ_3 is the rotation Angle of coordinate system {3} about axis y_2, z_5 is the distance that coordinate system {5} moves along axis z_4, θ_6 is the rotation Angle of coordinate system {6} about axis y_5. z_1, θ_2, z_5, θ_6 correspond with the actual quantity of each joint manipulator, and θ_3 with actual manipulator is displacement of the curved pipe s corresponding physical quantities, their conversion relations as follows:

$$ \theta_3 = \frac{s}{R} \tag{9.3.28} $$

where, R is the bending radius of the bending tube and is a constant value.

If the mark $c\theta_i = \cos\theta_i$, $s\theta_i = \sin\theta_i$, the expression of ${}^{0}_{6}T$ can be obtained as follows:

$$ {}^{0}_{6}T = {}^{0}_{1}T^{1}_{2}T^{2}_{3}T^{3}_{4}T^{4}_{5}T^{5}_{6}\,T = \begin{bmatrix} n_x & o_x & a_x & p_x \\ n_y & o_y & a_y & p_y \\ n_z & o_z & a_z & p_z \\ 0 & 0 & 0 & 1 \end{bmatrix} \tag{9.3.29} $$

where:

$$ \begin{aligned} n_x &= c\theta_2 c\theta_3 c\theta_6 - s\theta_2 s\theta_6 + e_1 \\ n_y &= s\theta_2 c\theta_3 c\theta_6 + c\theta_2 s\theta_6 \\ n_z &= -s\theta_3 c\theta_6 \\ o_x &= -c\theta_2 c\theta_3 s\theta_6 - s\theta_2 c\theta_6 \\ o_y &= -s\theta_2 c\theta_3 s\theta_6 + c\theta_2 c\theta_6 \\ o_z &= s\theta_3 s\theta_6 \\ a_x &= c\theta_2 s\theta_3 \\ a_y &= s\theta_2 s\theta_3 \\ a_z &= c\theta_3 \\ p_x &= c\theta_2 s\theta_3 z_5 + Rc\theta_2 (1 - c\theta_3) \\ p_y &= s\theta_2 s\theta_3 z_5 + Rs\theta_2 (1 - c\theta_3) \\ p_z &= c\theta_3 z_5 + Rs\theta_3 + z_1 \end{aligned} \tag{9.3.30} $$

9.3.4.3 Inverse Kinematics Analysis

In the previous section, we have analyzed the forward kinematics of the manipulator, but in the practical application of the manipulator, people are often more concerned about the inverse kinematics of the manipulator. Therefore, on the basis of the forward kinematics analysis, this section will discuss the inverse kinematics of the manipulator. When the expected position and attitude of the actuator connecting rod in the Cartesian coordinate system are known, how to calculate the corresponding meet the requirements of joint variables, or calculate the number of $z_1, \theta_2, \theta_3, z_5, \theta_6$ in the case of ${}_6^0T$. According to the working environment of the manipulator, when the actuator at the end of the manipulator coincides with the normal direction of the inner wall of the organ, it is in the best operating position. That is, once the location of the diseased tissue is determined, the position and axis direction of the end-effector can be determined as the inner wall of the organ is expanded into a fixed shape by the stretching device. The rotation Angle of the end-effector around the z_6 axis can be adjusted by visual observation to obtain the best operating attitude. Therefore, when the inverse kinematics of the manipulator is analyzed in this section, the degree of freedom of end-effector rotation around the z_6 axis is not considered, that is, ${}_6^5T$ coordinate transformation is omitted.

By inverting the two ends of $\left[{}_1^0T\right]^{-1}$ multiplied by Eq. (9.3.26), we get:

$$\left[{}_1^0T\right]^{-1}{}_5^0T = {}_2^1T{}_3^2T{}_4^3T{}_5^4T \tag{9.3.31}$$

Substituting the data, we can get:

$$\begin{bmatrix} n_x & o_x & a_x & p_x \\ n_y & o_y & a_y & p_y \\ n_z & o_z & a_z & p_z - z_1 \\ 0 & 0 & 0 & 1 \end{bmatrix} = \begin{bmatrix} c\theta_2 c\theta_3 & -s\theta_2 & c\theta_2 s\theta_3 & c\theta_2 s\theta_3 z_5 + c\theta_2 R\,(1 - c\theta_3) + e_1 \\ s\theta_2 c\theta_3 & c\theta_2 & s\theta_2 s\theta_3 & s\theta_2 s\theta_3 z_5 + s\theta_2 R\,(1 - c\theta_3) \\ -s\theta_3 & 0 & c\theta_3 & c\theta_3 z_5 + Rs\theta_3 \\ 0 & 0 & 0 & 1 \end{bmatrix} \tag{9.3.32}$$

If the corresponding terms of the matrix are equal, the equations can be obtained:

$$\begin{cases} o_y = \cos\theta_2 \\ n_z = -\sin\theta_3 \\ p_x = z_5\cos\theta_2\sin\theta_3 + R\cos\theta_2\,(1 - \cos\theta_3) + e_1 \\ p_z - z_1 = z_5\cos\theta_3 + R\sin\theta_3 \end{cases} \tag{9.3.33}$$

In formula $\theta_2 \in \begin{cases} [0, \pi/2), \left(p_x > 0,\ p_y \geq 0\right) \\ [\pi/2, \pi), \left(p_x \leq 0,\ p_y > 0\right) \\ [\pi, 3\pi/2), \left(p_x < 0,\ p_y \leq 0\right) \\ [3\pi/2, 2\pi), \left(p_x \geq 0,\ p_y < 0\right) \end{cases}$; $\theta_3 \in (0, \pi/2)$; $z_1 \in (0, 20)$; $z_5 \in (0, 20)$

Some unreasonable solutions can be eliminated according to the range of parameters of these joint variables, and the equation can be solved as follows:

$$\begin{cases} \theta_2 = \arccos\left(o_y\right) \\ \theta_3 = \arcsin\,(-n_z) \\ z_5 = \frac{Ro_y(1-a_z)+e_1-p_x}{o_y n_z} \\ z_1 = p_z - a_z z_5 + Rn_z \end{cases} \tag{9.3.34}$$

9.3.4.4 Workspace Analysis

Workspace is important in evaluating the robot performance, it represents the working area that the end-effector can reach. The workspace of the manipulator refers to the working space of the slave operator, that is, the range that the end clamp can operate after the manipulator reaches the working place. This space is also called the reachable space or total working space, denoted as W(P). This relationship can be regarded as the mapping of joint space variables to workspace, expressed as:

$$W(P) = \{P(q) : q \in Q\} \subset \Re^3 \tag{9.3.35}$$

In the formula: $P(q) : Q \rightarrow \Re^3$ is the positional component of the positive solution of the kinematic mapping, q is the generalized joint variable, Q is joint space, W(P) is the work space, $\Re^3$ is three-dimensional space.

When the structure parameters of the manipulator are given, the workspace is determined by the generalized coordinate $q_i = [q_1, q_2 \ldots q_n]$. Due to the limited institutions and driving devices, each joint variable has a certain range, $q_i^L \leq q_i \leq q_i^U$, $i = 1, 2, \ldots,$ n, work space can be represented as:

$$w_R = \begin{cases} x_p(q_i) \\ y_p(q_i) \\ z_p(q_i) \end{cases} \left(q_i^L \leq q_i \leq q_i^U\right) \tag{9.3.36}$$

Thus the workspace at the end of the operator can be obtained:

$$p = \begin{cases} x(q) = z_5 c\theta_2 s\theta_3 - Rc\theta_2 c\theta_3 + Rc\theta_2 \\ y(q) = z_5 s\theta_2 s\theta_3 - Rs\theta_2 c\theta_3 + Rs\theta_2 \\ z(q) = z_5 c\theta_3 + Rs\theta_3 + z_1 \end{cases} \tag{9.3.37}$$

Being limited by the structure and the working conditions, each joint variable range is as follows: $z_1 \in [0, 30\text{ mm}]$, $z_5 \in [0, 30\text{ mm}]$, $\theta_2 \in [0, 2\pi]$, $\theta_3 \in [0, \pi/2]$. Structural parameter: $R = 15$ mm.

MATLAB software was used for simulation analysis of the workspace. The following is the simulation result of the workspace. Figure 9.31 describes the reachable workspace formed by the reference point of the end of the manipulator under the base coordinates.

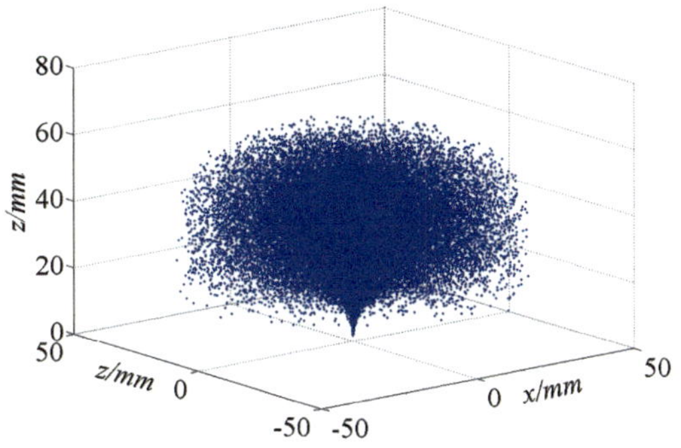

Fig. 9.31 Operator end Monte Carlo workspace [27]

9.4 Tendon-Sheath Transmitted Scale to Scale Mapping Remote Operation Manipulator

The purpose of master-slave mapping control is to replicate the actions of the master operator to the slave terminal of the system, thereby achieving real-time pose tracking of the master-slave operation, and consequently enhancing the operator's sense of immersion during surgical procedures. Due to the requirement of a large workspace for surgical robotic arms to meet the demands of flexible surgical operations, we propose a proportional mapping operator. This operator allows the small-scale movements of the human hand to control the large-scale movements of the robotic arm, and

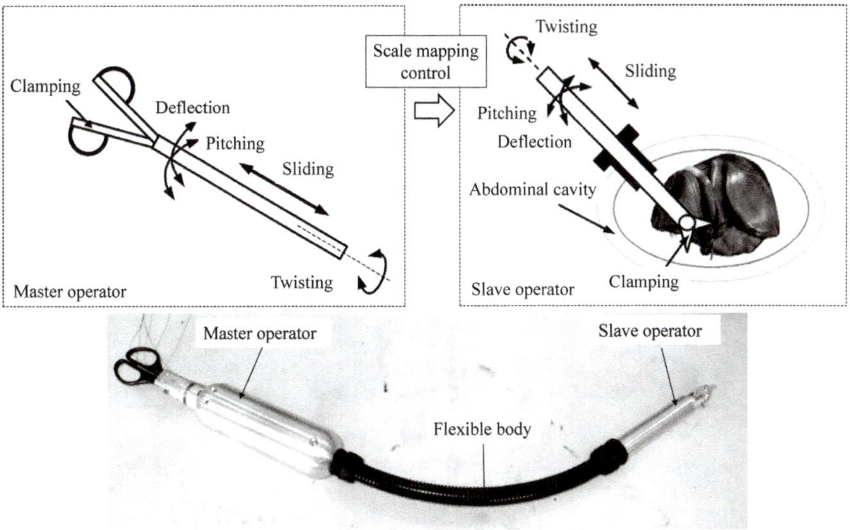

Fig. 9.32 The tendon-sheath actuated scale mapping operator [26]

during the motion process, both entities exhibit a certain proportional relationship in their motion and posture changes [26].

The manipulator device is mainly composed of three parts, the master operator, the flexible body and the slave operator, as Fig. 9.32 shows. Both the master operator and the slave operator have 5 degrees of freedom, which are respectively clamping, deflection, pitching, twisting and the sliding for moving forward and backward. There is a one-to-one correspondence between the corresponding degrees of freedom. The flexible body is hollow hose, mainly used to passively adapt to the bending complex environment. The tendon-sheath mechanism passes through the flexible body and remotely transmits the motion of the moving joints of the master operating box to the slave operator. At the same time, the tendon-sheath transmission mechanism also plays a role in increasing the stiffness of the flexible body.

9.4.1 Slave Operator Design

The three-dimensional structure diagram is shown in Fig. 9.33. The slave operator is mainly composed of clamping, deflection joint of two degrees of freedom, pitching motion joint, twisting joint rotating around the spindle and sliding joint moving along the spindle center line. Among them, the parts of twisting joint and prismatic joint are covered in the shell of the slave operator to improve the appearance of the whole mechanism. Obviously, the maximum external size of the slave operator depends on the diameter of the shell of the actuator. In order to meet the requirements of the design of the manipulator, the diameter of the shell of the slave operator is limited to 26 mm by constantly optimizing the structure of each motion joint of the slave operator. The size can basically meet the general surgical standards. In addition, a trimmer potentiometer is used to measure the Angle of each revolute joint, and a bellows joint is used to connect the slave operator to the intermediate flexible body.

The tendon-sheath transmission path for each joint of the slave operator are presented in Fig. 9.34. The sliding joint is driven by a single tendon-sheath system, it is used to adjust the position of the gripper in the direction of the spindle axis of the

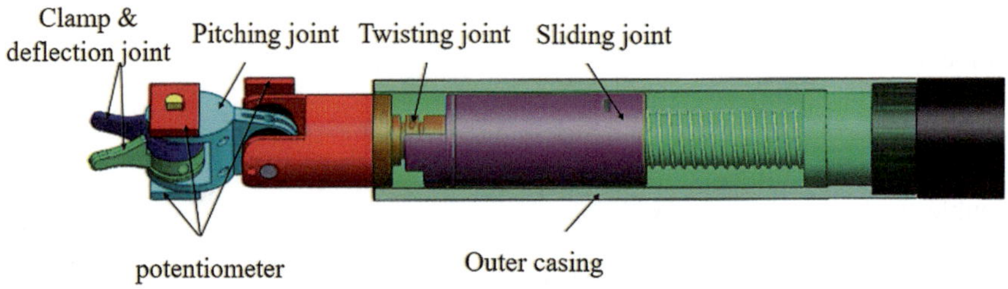

Fig. 9.33 3D view of the slave operator [26]

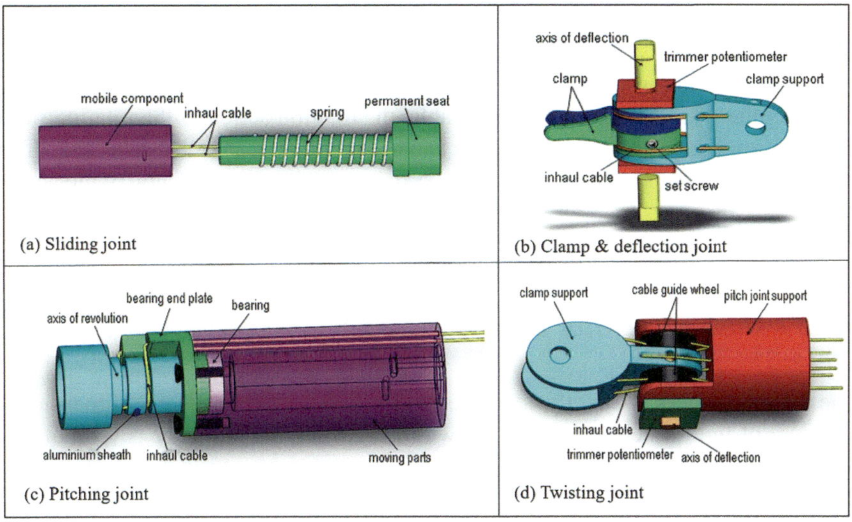

Fig. 9.34 Tendon-sheath transmission path for each joint of the slave operator [26]

slave operator. The input of the driving end is mainly transmitted to the moving part of the slave operator through two cables. When the cables are pulled, the moving parts slide forward, and the spring outside the convex shaft is compressed; when the cable is released, the recovery force of the spring can drive the moving part back, so that the straight prismatic joint can move forward and backward. The joint movement distance is 8 ± 1 mm.

The clamping and deflecting joints are used to realize the left and right clamping actions on specified objects. As shown in Fig. 9.34b, the two clamps of the joint move independently, each clamp is connected with its deflection shaft and fine tuning potentiometer, and then together with the support to form a rotation pair. The rotation range of the clamp is $\pm 129°$. Through the coordination between the two clamps, the joint can realize the clamping action on the specified object at any position within the rotation range of the clamp. The joint is driven by double tendon-sheath mechanism, and the end of the cable is fixed in the slot of the clamp. When the driving device pulls the cable, the clamp rotates around the respective deflection axis, and the angle of the deflection axis can be measured by the fine-tuning potentiometer.

The pitching joint is used to rotate the gripper around the pitching joint support, as shown in Fig. 9.34c. Similar to the deflection joint, a rotating pair is formed between the clamp support of the joint and the pitch joint support. The joint is driven by double tendon-sheath system, and the end of the cable is fixed in the groove of the clamp support. When the cable is pulled, the clamp support rotates around the deflection axis. The angle of the deflection axis can be measured by the fine-tuning potentiometer, and the angle range is $\pm 90°$.

The twisting joint is used to realize the rotation of the gripper around the spindle of the slave operator. As shown in Fig. 9.34d. The cable driving the joint with a spatial vertical configuration, and its motion principle is similar to that of bevel gear. By changing the winding direction of the cable, the rotation action of the rotating shaft is transformed into the straight forward and backward motion of the two cable. The joint was driven by double tendon-sheath system, and the rotation angle of the whole joint was ±180°.

9.4.2 Master Operator Design

The master operator is directly controlled by the human hand, the three-dimensional structure of the master operator is shown in Fig. 9.35, the diameter is 60 mm and the length is 145 mm. The joint structure of the master operator is similar to that of the corresponding joint structure of the slave operator. In this way, when the doctor operates the master operator, the slave operator inserted into the patient can copy the movements of the doctor's hand, which is equivalent to the doctor directly operating the joint movement of the slave operator, which can increase the doctor's immersion in the operation process and greatly reduce the difficulty of operation.

The master operator is also composed of four joints, including clamping, deflection joint, pitching joint, twisting joint and sliding joint, and the structure of each joint is similar to that of the corresponding joint of the slave operator. The manipulator can realize complex and diverse operations similar to human hands. The front end of the operation box is connected with the middle flexible body of the manipulator through the bellows joint, and the end is connected with a pair of scissors to drive the clamping and deflection joints. The fine-tuning potentiometer and three-axis load force acceleration sensor determine the precise rotation angle of each joint, while the movement of the prismatic joint is gauged using a scale. With these measurement tools, the operator can visually track changes in movement parameters of each joint from the control box, such that to better control the manipulator.

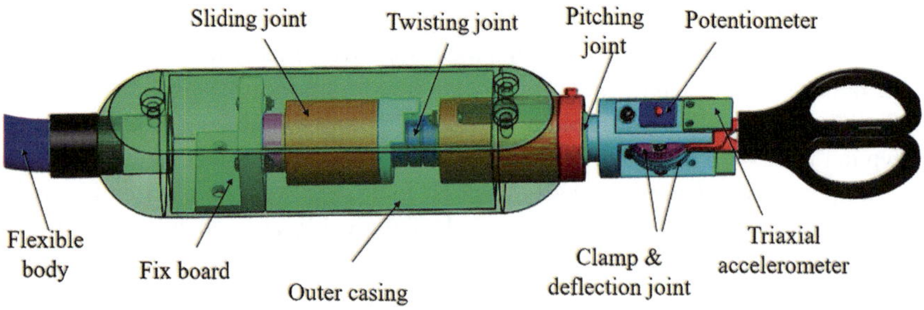

Fig. 9.35 Tendon-sheath transmission path for each joint of the master operator [26]

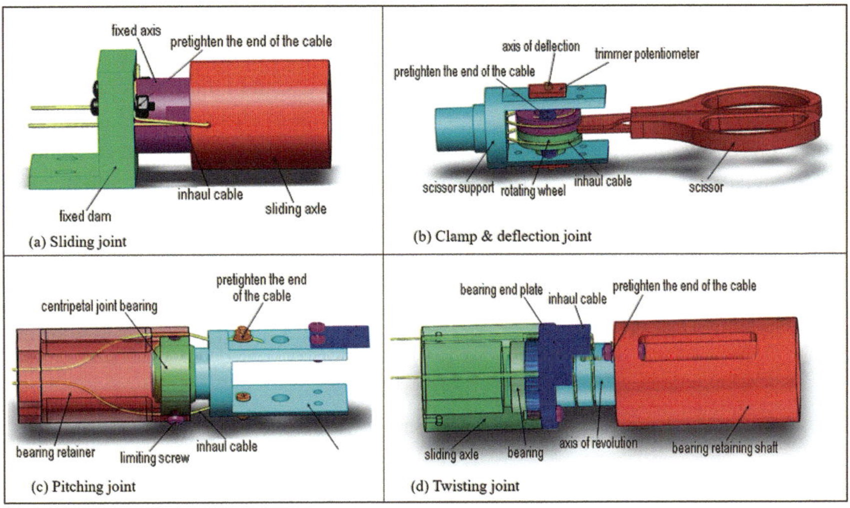

Fig. 9.36 Three-dimensional diagram of the master operator [26]

The structure and the tendon-sheath transmission path of these joints are shown in Fig. 9.36. The structure of the sliding joint of the master operator is shown in Fig. 9.36a. It is similar to that of the slave operator, and the size of this part of the structure is about twice that of the slave operator. Different from the slave operator, the cable end of the joint is fixed in a movable manner and is pretightened by a cable pretightening device similar to the pretightening device for bicycle brake lines.

The clamping and deflection joint structure of the master operator is basically the same as that of the slave operator. The clamping action of the clamping device at any position within the rotation range is realized through the coordination of the two clamping devices. The three-dimensional diagram of the joint is shown in Fig. 9.36b. The gripper of the joint is directly controlled by the human hand. In the operation process, the human hand directly controls the clamping and rotation of the scissors to make the slave operator clamp realize the corresponding action.

The pitching joint is different from that of the slave operator. Instead of the common deflection joint structure, centripetal joint bearings are used for this part of the structure, as shown in Fig. 9.36c. The angle sensor on the scissors support is a three-axis acceleration sensor, which is used to measure the rotation Angle of the overturning joint and the pitching joint. As can be seen from the three-dimensional figure, when the scissors support rotates up and down with the inner ring of the ball bearing for a certain angle, the scissors support will contact with the end face of the fixed shaft of the ball bearing and cannot continue to rotate. Through calculation, the upper and lower rotation range of the scissors support is about ±24°, without considering the displacement loss of the tendon-sheath in the transmission process, through theoretical calculation, when the scissors support rotates to the maximum Angle, the maximum rotation Angle of the pitching joint of the slave operator can reach ±62°, which can meet the movement requirements of the general minimally invasive surgery for the manipulator.

The twisting joint of the master operator is similar to that of the slave operator, and this part of the structure is also about twice size of that of slave operator. According to theoretical calculation, without considering the displacement loss in the tendon-sheath transmission process, as long as the part of the structure rotates about ±89°, the turnover joint of the slave operator can realize the rotation range of ±180°. Here, the fixed shaft of ball bearing rotates along with the rotating shaft through the thread matching with the rotating shaft, and the two stay cables that control the rotation of the rotating shaft pass through the central hole of the rotating shaft and are fixed on the fixed shaft of ball bearing by the stay cable pretightening device, as shown in Fig. 9.36d.

9.4.3 The Passive Flexible Body

The slave operator and master operator communicate motion through a tendon-sheath drive system, where a tightly wound spring sleeve serves as the flexible framework of the robot. This framework is encased in a plastic bellows, connected to both operators via bellows joints, as depicted in Fig. 9.37. While the plastic bellows offer good bending capabilities, they lack sufficient axial stiffness, making them prone to jamming in tight spaces. Incorporating a spring sleeve enhances the axial stiffness. Moreover, because the spring sleeve can bend sideways, the combined flexibility of the spring sleeve and bellows allows for better conforming movement.

9.4.4 Scale Mapping Control

The manipulator control operates on a master-slave system. Using the main arm (master operator) and the tendon-sheath transmission, one can direct the movements

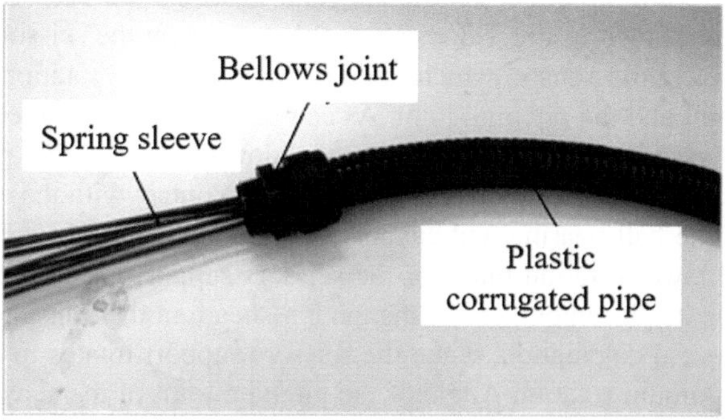

Fig. 9.37 Manipulator passive flexible body [26]

of the follower arm (slave operator). During the operation, doctors only need to control the movements of each joint of the master operator(d_1', β_2, β_3, β_4), and then the system can map these movements proportionally to the corresponding joints of the slave operator(d_1, θ_2, θ_3, θ_4). Here, d_1', β_2, β_3, β_4 are the motion parameters of the straight moving joint, the rollover joint, the pitching joint and the deflection joint of the master operator, respectively, and d_1, θ_2, θ_3, θ_4 are the motion parameters of the linear mobile joint, flipping joint, pitching joint and deflection joint of the slave operator respectively.

9.4.4.1 Theoretical Mapping Ratio

In proportional mapping control, the expansion and contraction in the tendon-sheath transmission process are disregarded. This allows for a theoretical deduction of the proportional relationship between the motion parameters of each joint on the slave operator and its corresponding joint on the master operator. The d_1 and d_1', θ_2 and β_2, θ_3 and β_3, θ_4 and β_4 the ratio between the relationship.

For linear movement joints, the slave operator displacement d_1 and tail box operation displacement d_1' are equal. For the rollover joint, the rollover shaft radius of the slave operator is $r_2 = 4$ mm, and the rollover shaft radius of the master operator is $R_2 = 8.25$ mm. Considering the wire diameter of the cable, it can be obtained: $r_2/R_2 = (4 + 0.25)/(8.25 + 0.25) = 1 : 2$, so $\theta_2/\beta_2 = R_2/r_2 = 2$. Similarly, for the deflection joint, the deflection shaft radius of the slave operator is $r_4 = 5.75$ mm, and the half-diameter of the deflection shaft of the master operator is $R_4 = 11.75$ mm. Considering the wire diameter of the cable, it can be obtained: $r_4/R_4 = (5.75 + 0.25)/(11.75 + 0.25) = 1 : 2$, so $\theta_4/\beta_4 = R_4/r_4 = 2$. For longitudinal joints, the slave operator pitching axis radius $r_3 = 5.41$ mm, the tail of the operation box pitch joint USES is ball bearing structure, because of the pull around line direction is not the circumference direction of the ball bearing, so the pitch of the pitch Angle joints before and after θ_3 is not a linear relationship with the β_3, through the theoretical derivation and calculation, θ_3 and β_3 relationship can be obtained as shown in Fig. 9.38.

Due to the operation restrictions of various parts in the longitudinal joint of the tail box, the maximum value of β_3 is 25.21°. It can be seen from Fig. 9.38, when β_3 changes within the range of 0~18°, θ_3 increases rapidly, with the increase of β_3 is almost linear relationship, the ratio of θ_4/β_4 probably remain at around 3; When β_3 in 18°~23.76° with the increase of β_3, θ_3 amplitude decreases, and when $\beta_3 = 23.76°$, θ_3 reached a maximum of 62.13°; When β_3 over 23.76° until it reached a maximum of 25.21°, θ_3 decreases as the β_3 instead. Therefore, in order to ensure the continuity of the operation, operation box pitch joint movement at the end of the operation, should try to make β_3 within ±23.76° movement, within the scope of the Angle of the slave operator longitudinal joints can reach the maximum pitching Angle as ±62.13°.

Table 9.2 The proportional relationship between the corresponding joint parameter changes of the slave operator and the master operator

	$d_1 : d_1'$	$\theta_2 : \beta_2$	$\theta_3 : \beta_3$	$\theta_4 : \beta_4$
Proportionality	1 : 1	2 : 1	(Fig. 9.38)	2 : 1

Through theoretical calculation, we have deduced the d_1 and d_1', θ_2 and β_2, θ_3 and β_3, θ_4 and β_4 (Table 9.2), the ratio between the relationship, in this way, in the process of surgery, doctors only observe the change of hand movement parameters, through scale conversion, can learn the head of the manipulator actuator corresponding movement parameters change; Conversely, after the motion path of the manipulator's slave operator is planned, the size of the motion parameters of the human hand can be deduced by proportion conversion, so as to guide the doctor to complete the corresponding surgical operation.

9.4.4.2 Actual Mapping Ratio

In practice, the corresponding proportion relationship between parameter may be influenced by other factors, so we need to calculate the actual scale mapping relationship, thus to proportion relations make appropriate modifications to the theory.

1. The proportionality verification of d_1 and d_1'
   We attach a scale label to the sliding block to measure its movement. In experiments, starting from the initial position, we move it forward by 1 mm increments,

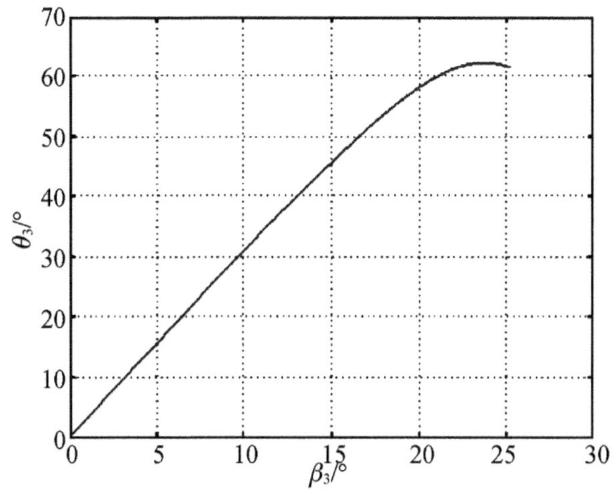

Fig. 9.38 The theoretical proportional relationship between θ_3 and β_3 [26]

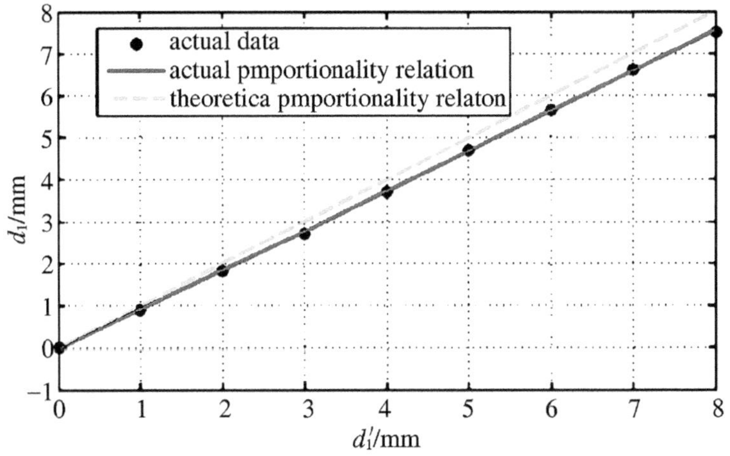

Fig. 9.39 The actual and theoretical proportional relations between d_1 and d_1' [26]

recording values as d_1' and d_1, With a maximum movement range of 8 mm, we collect a set of experimental data, as shown in Fig. 9.39. Such that we can get an actual proportional relationship between d_1 and d_1' by analyzing and fitting this data While there's some deviation from the theoretical proportional curve, the overall shape remains consistent. This suggests that the mapping effect for the tendon-sheath-driven linear prismatic joint motion is quite effective.

2. The proportionality verification of θ_2 and β_2
For the rollover joints in both the slave and master operators, we use a triaxial acceleration sensor to measure the joint angle. This sensor measures joint angles based on acceleration principles, which work as spatial vectors. It's essential because it captures an the object motion across three axes and can determine motion direction when it's unknown. Since the triaxial acceleration sensor is based on the principle of gravity, it can be used to measure the dip angle of biaxial ±90° or biaxial 0~360°.

After calibration of triaxial acceleration sensor module, we conduct experiment to scale mapping θ_2 and β_2 mapping relationship. We observed that when β_2 is less than 30°, the increase in β_2 results in a slow growth of θ_2. However, once β_2 over 30°, θ_2 growth significantly accelerates. This suggests that in the tendon-sheath transmission model with flipped joints, initially, the output changes only slightly with increasing input. There's a noticeable delay in the actual output compared to the theoretical curve, especially within this delay zone where the tendon-sheath transmission displays more nonlinear behavior and greater displacement loss. However, as we move beyond this zone, the tendon-sheath transmission becomes more linear, approaching the desired mapping effect.

In fact, the errors of the theoretical results and the experimental results in the scale mapping relations between d_1 and d_1', θ_2 and β_2, θ_3 and β_3, θ_4 and β_4 mainly come from two main sources.

(1) Nonlinear tendon-sheath transmission
Since the revolute joints of the manipulator are driven by double tendon-sheath mechanism, there is a certain displacement dead zone happened at starting, stretching, and relaxing phases. it greatly influence the motion mapping relationship between the slave operator and the corresponding revolute joints of the master operator.
(2) Joint Interference
When one joint moves, nearby joints can unintentionally shift because of the shared tendon-sheath. This interference leads to initial movements in affected joints, causing mapping errors in their motion.

To address these issues, we can apply an initial preload to the cable, using lubricants to lessen friction, and optimizing the transmission route to reduce sharp bends. Additionally, we can strategically arrange the tendon-sheath transmission paths to ensure independent movement of each joint without affecting adjacent ones.

Additionally, there are many other factors affecting the motion mapping relationship between the slave operator and the master operator. Therefore, it is necessary to optimize the structure of the manipulator and further study the tendon-sheath transmission characteristics. Based on the experimental results, we can better understand the manipulator's movements. Moreover, it helps doctors anticipate and execute their hand movements, and enhance the overall operation effectiveness.

References

1. Le H, Do T, Phee S (2016) A survey on actuators-driven surgical robots. Sens Actuators, A 247:323–354
2. Lu V, Kerezoudis P, Gilder H et al (2017) Minimally invasive surgery versus open surgery spinal fusion for spondylolisthesis: a systematic review and meta-analysis. Spine 42(3):E177–E185
3. Sackier J, Wang Y (1994) Robotically assisted laparoscopic surgery: from concept to development. Surg Endosc 8:63–66
4. Boyd W, Kiaii B, Kodera K et al (2002) Early experience with robotically assisted internal thoracic artery harvest. Surg Lap Endosc Percut Tech 12(1):52–57
5. Pilecki M, McGuire B, Jain U et al (2014) National multi-institutional comparison of 30-day postoperative complication and readmission rates between open retropubic radical prostatectomy and robot-assisted laparoscopic prostatectomy using NSQIP. J Endourol 28(4):430–436
6. Lum M, Friedman D, Sankaranarayanan G et al (2009) The RAVEN: design and validation of a telesurgery system. Int J Robot Res 28(9):1183–1197
7. Curcillo P, Wu A, Podolsky E et al (2010) Single-port-access (SPA TM) cholecystectomy: a multi-institutional report of the first 297 cases. Surg Endosc 24:1854–1860
8. Yao H, Ngu J, Lin Y et al (2020) Robotic transanal minimally invasive surgery for rectal lesions. Surg Innov 27(2):181–186

9. Kalloo A, Singh V, Jagannath S et al (2004) Flexible transgastric peritoneoscopy: a novel approach to diagnostic and therapeutic interventions in the peritoneal cavity. Gastrointest Endosc 60(1):114–117
10. Marescaux J, Dallemagne B, Perretta S et al (2007) Surgery without scars: report of transluminal cholecystectomy in a human being. Arch Surg 142(9):823–826
11. Tortora G, Salerno M, Ranzani T et al (2013) A modular magnetic platform for natural orifice transluminal endoscopic surgery. In: 2013 35th annual international conference of the IEEE engineering in medicine and biology society (EMBC). IEEE, pp 6265–6268
12. Lehman A, Wood N, Dumpert J et al (2008) Dexterous miniature in vivo robot for NOTES. In: 2008 2nd IEEE RAS & EMBS international conference on biomedical robotics and biomechatronics. IEEE, pp 244–249
13. Abbott D, Becke C, Rothstein R et al (2007) Design of an endoluminal NOTES robotic system. In: 2007 IEEE/RSJ international conference on intelligent robots and systems. IEEE, pp 410–416
14. Lee J, Shin W, Kwon D (2013) A modularized wire-driven overtube using tube type bending mechanism for robotic NOTES system. In: 2013 13th international conference on control, automation and systems (ICCAS 2013). IEEE, pp 631–634
15. Francis P, Eastwood K, Bodani V et al (2018) Design, modelling and teleoperation of a 2 mm diameter compliant instrument for the da Vinci platform. Ann Biomed Eng 46:1437–1449
16. Wang H, Zhang R, Chen W et al (2017) A cable-driven soft robot surgical system for cardiothoracic endoscopic surgery: preclinical tests in animals. Surg Endosc 31:3152–3158
17. Bernth J, Arezzo A, Liu H (2017) A novel robotic meshworm with segment-bending anchoring for colonoscopy. IEEE Robot Autom Lett 2(3):1718–1724
18. Abidi H, Gerboni G, Brancadoro M et al (2018) Highly dexterous 2-module soft robot for intra-organ navigation in minimally invasive surgery. Int J Med Robot Comput Assist Surg 14(1):e1875
19. Luo K, Looi T, Sabetian S et al (2018) Designing concentric tube manipulators for stability using topology optimization. In: 2018 IEEE/RSJ international conference on intelligent robots and systems (IROS). IEEE, pp 1764–1769
20. Dwyer G, Chadebecq F, Amo M et al (2017) A continuum robot and control interface for surgical assist in fetoscopic interventions. IEEE Robot Autom Lett 2(3):1656–1663
21. Burgner J, Rucker D, Choset H (2015) Continuum robots for medical applications: a survey. IEEE Trans Rob 31(6):1261–1280
22. Runciman M, Darzi A, Mylonas G (2019) Soft robotics in minimally invasive surgery. Soft Rob 6(4):423–443
23. Frede T, Stock C, Renner C et al (1999) Geometry of laparoscopic suturing and knotting techniques. J Endourol 13(3):191–198
24. Cleary K (2006) Medical robotics and the operating room of the future. In: 2005 IEEE engineering in medicine and biology 27th annual conference. IEEE, pp 7250–7253
25. Shen D, Ma K, Tian M et al (2019) A lariat-based dilatation device for hysteroscopy: an in vitro study. Ann Transl Med 7(18)
26. Yuan P (2014) Design of a tendon-sheath-driven manipulator with proportional mapping control. Southeast University
27. Shen D (2015) The research of expansion and operation technology for endovascular surgical robot. Southeast University

Chapter 10
Tendon-Sheath Transmitted Soft Light Exoskeletons

Abstract Exoskeletons is a type of wearable robot system working in parallel with human limbs and combining strength of robot with human intelligence. It can effectively improve human strength, endurance, and locomotion in various environments. In the last decade, many researchers have developed different kinds of wearable exoskeleton for the disabled, elderly, and soldiers. These devices are equipped with different types of actuators, such as electrical motors, hydraulic cylinders and artificial muscles, and thus able to assist human bodies to move with any assistance level of training. Currently, light weight, portability and high driving capability are the main directions for the development of exoskeletons. To this end, we took advantage of the high compliance, lightweight and large transmission force of the tendon-sheath mechanism to develop wearable exoskeletons. In this chapter, we introduce several kinds of wearable exoskeletons actuated by the tendon-sheath systems, including the upper limb and the lower limb exoskeletons. The structure design, movement control of the tendon-sheath actuated exoskeletons were illustrated, and the tendon-sheath based artificial muscle and its application in exoskeleton is also introduced. We hope that through this chapter, readers can understand the general steps of using tendon-sheath mechanism to develop exoskeleton systems, and further inspire readers to expand the application of tendon-sheath in other robotic systems.

10.1 State of the Art on Lightweight Exoskeleton

Humans have always desired to possess strong physical strength or to restore the ability to walk for those who have difficulty walking. Lower limb exoskeletons, as a type of parallel wearable bipedal robot attached to the sides of the human legs, are mechanical devices created by humans to fulfill these dreams. This category of mechanical devices, while synergistically moving with the human lower limbs, enhances limb strength and improves human mobility by providing assistive force to aid lower limb movement. In the field of engineering applications, lower extremity exoskeleton has a great application prospect, including assisting the elderly, weak, sick and disabled people to achieve walking and rehabilitation training, or improving the movement capacity of healthy people. Therefore, exoskeletons have attracted

X. Wang and Q. Zhang, *Tendon-Sheath Transmitted Light and Slender Robots*,
https://doi.org/10.1007/978-981-96-5581-6_10

extensive attention from scholars around the world, and various upper and lower extremity exoskeleton prototypes have been developed. Many prototypes have been applied to the human body and some of them are moving towards commercialization.

The history of exoskeletons can be traced back to the Hardiman exoskeleton jointly developed by General Electric Company and the U.S. military in the 1960s. The exoskeleton is driven by hydraulic and electric power, which can amplify the wearer's force by 25 times. The force feedback system can make the wearer perceive the gravity and operating status of lifting objects [1]. At the beginning of this century, the U.S. Defense Advanced Research Projects Agency (DARPA) launched the "exoskeleton to enhance human capabilities" project. The first phase of the project began in 2001, with the participation of the University of California, Berkeley, Sarcos Research and Oak Ridge National Laboratory. The second stage began in 2003, and Sarcos Research and Berkeley University of California were selected for analysis. The third stage began in 2004, and Sarcos Research Company is mainly engaged in the research of high-power upper and lower limb exoskeleton system with fast movement and heavy armor protection [2]. Subsequently, a wave of exoskeleton research and development has been triggered around the world. As a result, some famous exoskeletons have come out, such as the BLEEX developed by the UC Berkeley [3], the XOS launched by Sarcos Research Company [4], the ReWalk at Israel [5], etc.

However, these exoskeletons have one thing in common, that is, they are composed of a large number of rigid connecting rod structures, which are large in volume and mass and inconvenient to carry. In terms of driving, the rigid exoskeleton is mostly driven by motor, motor with reducer or hydraulic drive. Because of the large servo stiffness of motor drive, the comfort of human-machine interaction becomes poor, and it has certain risks. In addition, when using the motor drive mode, the motor is usually placed near the driven joint, which directly increases the motion inertia of the end of the limb. The hydraulic drive mode usually requires an external energy source, which directly leads to the poor portability of the exoskeleton, and the fluid modeling accuracy is not high, the control is difficult, and it is not easy to realize the personalized customization of the exoskeleton assistance.

In order to solve the limitations of rigid exoskeletons mentioned above, flexible exoskeletons have gradually become a research hotspot. Such exoskeletons use flexible materials (such as fabrics and ropes), flexible drives (such as pneumatic artificial muscles) or rigid materials for flexible structural design. Compared with the rigid exoskeleton described above, the flexible exoskeleton has many outstanding advantages, such as light weight, flexible driving mode, and the driver can be placed on the waist or back of the wearer to reduce the motion inertia of the lower extremity; Good flexibility, can adapt to the body shape and movement differences of the wearer; Because the assistance provided by the flexible exoskeleton is parallel to the relevant muscle groups of the human body and different from the rotation drive of the rigid exoskeleton to the joint, the flexible exoskeleton assistance is more natural; In addition, the flexible exoskeleton can match with daily shoes and clothing, reducing the psychological burden of the wearer when living with others. Based on the above advantages, flexible exoskeleton has high research value. The Wyss Bio-inspired Engineering Institute of Harvard University in the United States is at the forefront of the world in the field of flexible exoskeleton. Since 2013, the Wyss Laboratory Conor J. Walsh team has taken the lead in developing a new generation of flexible exoskeleton system (Soft Exosuit), which is mainly composed of flexible materials such as fabrics and elastomers[6]. The prototype is a multi-joint flexible lower limb exoskeleton actuated by the tendon-sheath mechanism, the total mass is only 10.1 kg.

In 2018, Wyss Lab improved the two-joint assisted exoskeleton to the outdoor version, with a total mass of 9.0 kg (including the power mass), which can provide 250N peak assistance for hip joint extension and 350N peak assistance for ankle metatarsal flexion and hip joint flexion, and the average metabolic energy of the wearer decreased by about 12.12%, about 0.59 W/kg [7]. In 2021, the team proposed an ankle exoskeleton that can provide customized assistance to the wearer. This exoskeleton adopts the muscle-based assistance strategy, and directly observes the activity of the soleus muscle of the lower leg through the ultrasonic imaging sensor, such that reduces the walking energy consumption of the soleus muscle [8].

In addition to the flexible exoskeleton research team of Wyss Laboratory of Harvard University, the Agrawal team of Roar Laboratory of Columbia University developed the lower limb exoskeleton system C-ALEX with flexible transmission function. The C-ALEX is also driven by tendon-sheath mechanism, the total mass of the exoskeleton is 4.3 kg, it can provide assistance for the flexion and extension of the hip and knee joints[9]. In addition to the tendon-sheath actuation, pneumatic artificial muscles are also adopted in soft exoskeletons [10].

These exoskeletons use motors, hydraulic devices and other power sources to drive human body movements, but the power devices are heavy, costly, and easily cause additional energy consumption. To realize more efficient walking ability, the unpowered exoskeletons inspired by the human walking system were emerged simultaneously. In 2015, Steven Collins team from the Biomechatronics Laboratory of Carnegie Mellon University in the United States developed an ankle joint unpowered exoskeleton [11]. The exoskeleton is composed of a spring, a ratchet mechanism, a lower leg and a sole frame made of lightweight carbon fiber materials, with a total weight of about 503 g. The mechanical ratchet mechanism is triggered by the angle of the ankle joint, and the swing phase of the lower limb and the support intersect to open and close, controlling the stretch energy storage and contraction energy release of the rear leg spring, so as to achieve the purpose of reducing the plantar flexion moment of the ankle joint and the activity of the calf muscle group. The experiment shows that the unpowered exoskeleton can reduce the metabolic energy consumption of the wearer by $7.2 \pm 2.6\%$ during the uniform speed walking on the flat and straight road [12].

However, the Collins non-powered ankle exoskeleton mentioned has some limitations. Its passive clutch mechanism doesn't adjust to changes in the wearer's gait or to outdoor conditions like uneven roads. In addressing this issue, they proposed an "active and unpowered exoskeletons", and the original clutch mechanism was replaced by a ratchet mechanism powered by a small motor. This motor is controlled by a microcontroller and reacts to pressure sensors at the heel's bottom [13]. Even though "active and unpowered exoskeleton" might sound confusing, it means that most of its energy goes into sensing, control, and managing the clutch. This design lets it function like an unpowered exoskeleton but uses less energy than a dynamic one.

In conclusion, the exoskeleton technology, as an interdisciplinary discipline, involves robotics, human biomechanics, mechanical mechanism, bionics, mechanical manufacturing, sensing and control engineering, signal processing and other disciplines. There are still several key challenges in the research of lower limb assist exoskeleton. Firstly, understanding how humans walk is vital, it ensures that the exoskeleton can move seamlessly with the wearer's intentions. Secondly, while efficiently powering the exoskeleton is essential, it also requires compact and adaptable technology capable of handling diverse outdoor conditions. Additionally, enhancing sensors and control systems becomes crucial for providing real-time feedback and ensuring smoother movements. Lastly, for these devices to gain widespread acceptance and usability, they must not only be lightweight and comfortable but also versatile enough to cater to various users and walking scenarios while maintaining safety standards.

10.2 Tendon-Sheath Transmitted Lower Limb Exoskeleton

Lower limb exoskeletons help with human walking by supporting the hip, knee, and ankle joints. As the human gait is regular and periodic, the lower limb exoskeletons usually worked based on the joint torque or displacement during the gait cycle. The exoskeletons that perform human-led torque assistance are called power-assisted exoskeletons, while the exoskeletons tracking joint trajectory are called rehabilitation exoskeleton. The power-assisted exoskeletons are usually used to assist soldiers, workers and other able-bodied people to reduce energy consumption during walk, while the rehabilitation exoskeletons are used for rehabilitation training to help stroke patients regain their walking ability.

10.2.1 Design Requirements for Exoskeleton

When designing the lower limb rehabilitation exoskeleton mechanism driven by the tendon-sheath mechanism, the following factors should be fully considered in the design process.

1. The patient can basically achieve such functions as sitting/standing up, walking with crutches when wearing the lower limb rehabilitation exoskeleton, such that the degree of freedom of the exoskeleton mechanism needs to meet the requirements of rehabilitation training and assistance.
2. The design should ensure the convenience of the installation and disassembly of the tendon-sheath device, and the tendon-sheath should have pre-tightening mechanism to ensure the transmission accuracy.

3. The motion range of each joint needs to be reasonably designed to make the exoskeleton mechanism more safe. It is necessary to set joint limitation position to prevent potential injuries from system failures or control errors.
4. The transmission path of the tendon-sheath should be smooth and the deformation curve should be minimum. The layout of the tendon-sheath path should be optimized to improve the efficiency of the tendon-sheath transmission, and then improve the efficiency of the tendon-sheath artificial muscle drive system.
5. Make the waist width of the exoskeleton adjustable and customize the leg pole lengths for different users. Once the length of the large and small leg poles changes, the corresponding optimal tendon-sheath layout path should change accordingly.
6. Realize the lightweight of structure design as much as possible, reduce the energy consumption of the system, and improve the endurance and wearing comfort of the mechanism.

10.2.2 Lower Limb Exoskeletons Actuated by Tendon-Sheath Mechanism

With the tendon-sheath mechanism, we developed several lower limb exoskeletons for single joints and whole lower extremity. The tendon-sheath actuated exoskeletons for single joint are shown in Fig. 10.1, including hip exoskeleton, knee exoskeleton, and ankle exoskeleton. These exoskeletons are all driven by double tendon-sheath mechanisms to realize the joint rotation in two directions. When the human body walks, the hip joint mainly works to flexion and extension, while its external and internal rotational motions are very small. Therefore, the hip joint of the exoskeleton in Fig. 10.1a mainly uses a pair of double tendon-sheath drive systems to assist the motion of flexion and extension, while the internal rotation and external rotation of the exoskeleton are set as passive degrees of freedom, which are passively adjusted according to the movement of the human body. Similarly, the knee exoskeleton in Fig. 10.1b is driven by a double tendon-sheath drive system to perform knee flexion and extension in one plane of motion. With the tendon-sheath mechanism, the drive motor can be placed on the waist or back, thus the motion inertia of the legs is reduced, such that the energy consumption is reduced. Moreover, it also ensures the light and suppleness of the knee exoskeleton. The ankle exoskeleton in Fig. 10.1c has two double tendon-sheath drive systems that responsible for the eversion/inversion and the dorsiflexion/planarflexion respectively. The use of tendon-sheath transmission mechanism can easily actuate the two degrees of freedom of the ankle, and ensure the lightness and simplicity of the structure of ankle exoskeleton.

By combining these single joint exoskeletons, a whole lower limb wearable exoskeleton can be obtained. However, the use of high-power motors and various linkage mechanisms makes the whole limb exoskeleton heavy and energy-intensive. To this end, we have conducted in-depth research on the energy saving and light weight of exoskeletons. Based on the human lower-limb movement mechanism and the

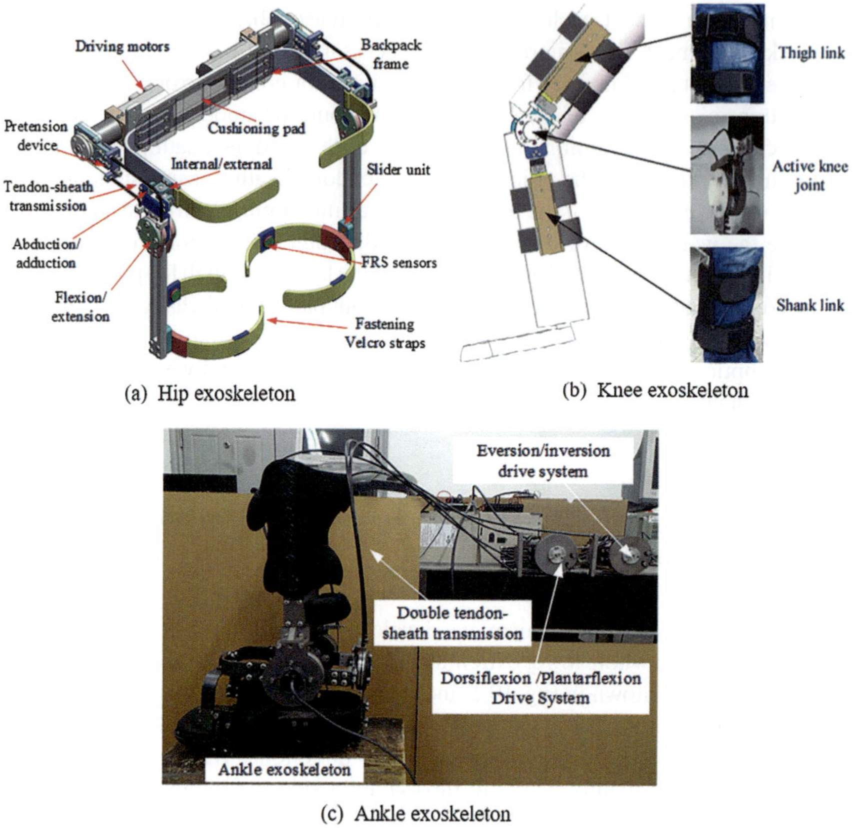

Fig. 10.1 The tendon-sheath actuated exoskeletons for single joint

passive bipedal walking theory, a quasi-passive power-assisted lower-limb exoskeleton driven by tendon-sheath actuation system is developed, shown in Fig. 10.2. The active hip joint is driven by a double tendon-sheath actuation system to accomplish the external placement of the joint drive motor and the far-end compliant driving. Such that the walking assistance is achieved by controlling the joint torque output. The passive compliant joint design is implemented by mechanical spring components to adjust the COM of the exoskeleton and comply the movement trend with the energy storage and release, aims to reduce the energy consumption and improve the mechanism performance. The impulse module along the thigh lever is used as the trigger signal for the man-machine initiative movement as pedal, plantar flexion, knee flexion and the thigh flexion to aid the accomplishment of the gait conversion and achievement of the lower limb strength training. The quasi-passive power-assisted lower-limb exoskeleton achieves lightweight structure, low energy consumption, and more coordination assistance performance.

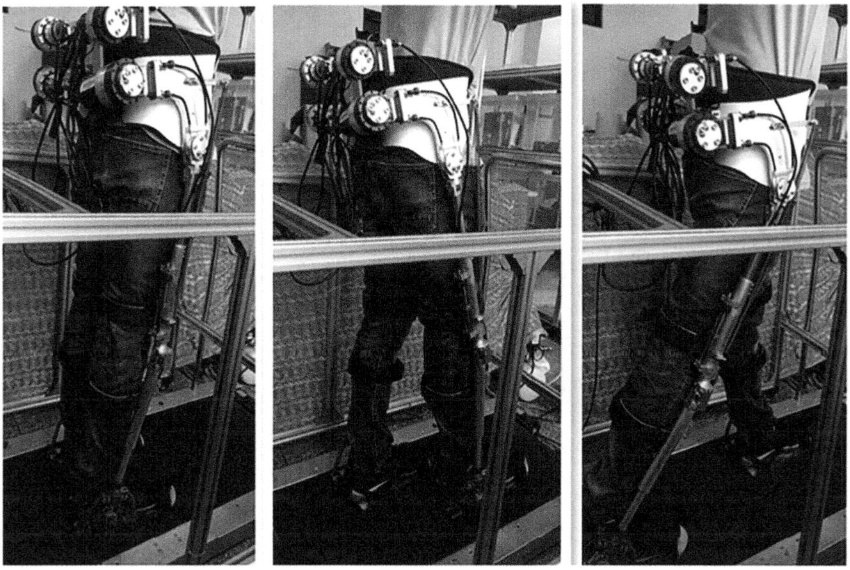

Fig. 10.2 The quasi-passive power-assisted lower-limb exoskeleton driven by tendon-sheath actuation system

10.2.3 Tendon-Sheath Artificial Muscle Actuated Exoskeleton

To further improve the wearable exoskeleton with bionic actuation performance, good portability and precise movement control, we propose a tendon-sheath artificial muscle based on the Hill muscle model and then develop a wearable exoskeleton. Artificial muscle has become a hot research topic in recent years with the technological development of smart materials and soft robots. They have controllable contractions, high energy density ratios, and soft and light structures. However, due to the deformation properties and the driving limitations of these materials, these artificial muscles are not so good for exoskeletons. Therefore, we use tendon-sheath mechanism to develop a more lightweight, controllable, and compliant artificial muscle actuator. The structural design, transmission characteristics analysis, and motion control strategies of the tendon-sheath artificial muscle and the control strategy for the exoskeleton are introduced here.

10.2.3.1 Compound Tendon-Sheath Artificial Muscle

In biological movement, skeletal muscles usually drive joint in the form of antagonist muscles, as Fig. 10.3a shows. The muscle that actively contracts during the joint movement is agonist muscle, while the relaxed or stretched muscle in the other side

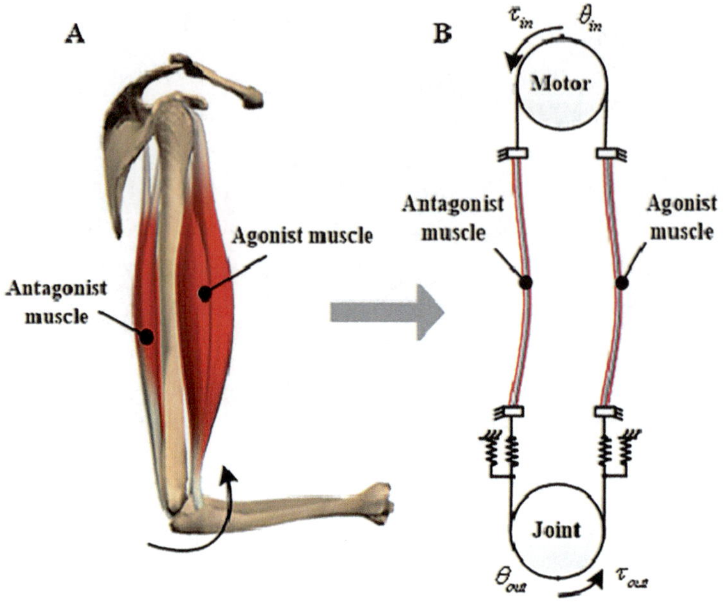

Fig. 10.3 The schematic diagram of compound tendon-sheath artificial muscle system [20]

is antagonist muscle. The agonist muscle plays a major role in joint movement, it provides positive force for joint movement. The antagonist muscle also plays an important role in the joint movement, it works in conjunction with agonist muscle to provide cushioning and protect the joint from excessive rotation. Muscle can be agonist muscle or antagonist muscle in different joint movements. Similarly, a compound tendon-sheath artificial muscle actuation system, shown in Fig. 10.3b, is proposed for imitating the movement of antagonist muscles. The two tendon-sheath artificial muscles are driven by one motor, and play the role of agonist muscle and antagonist muscle respectively.

10.2.3.2 Tendon-Sheath Artificial Muscle Actuated Exoskeleton

Using the compound tendon-sheath artificial muscle, we developed a lower limb rehabilitation exoskeleton system [14]. As Fig. 10.4 shows, the system includes a lower limb exoskeleton prototype, a motor driving set, a sensing module, a micro controller and a human-machine interface. The hip and knee joints of the lower limb exoskeleton are active joints that driven by compound tendon-sheath artificial muscles, while the external swing/adduction degrees of freedom of the hip joint and the ankle joint are all passive joints without actuation.

Since the tendon-sheath transmission mechanism can realize random curved transmission path, the four motors (Maxon DC servo brush motor RE50) used to drive

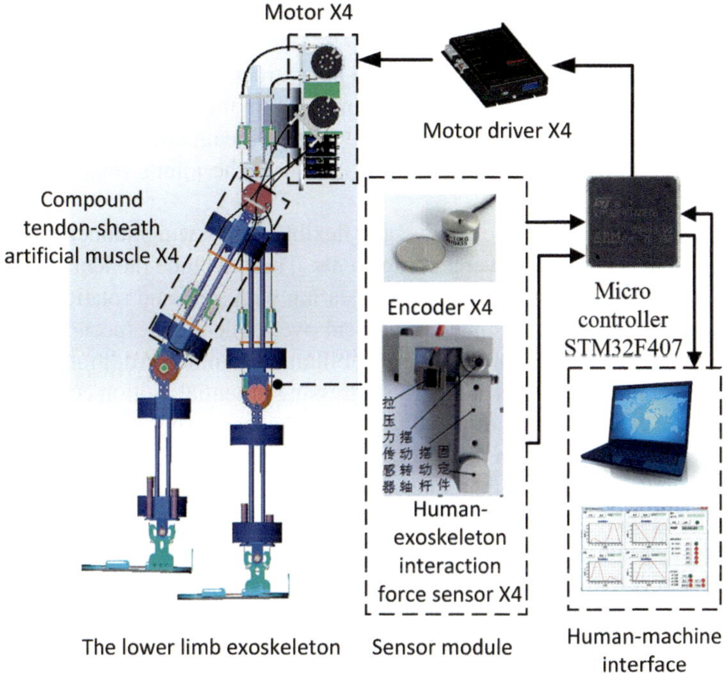

Fig. 10.4 The lower-limb exoskeleton system actuated by compound tendon-sheath artificial muscles [20]

the left and right hip and knee joints of the exoskeleton can be placed at the waist backpack. Thereby, the moving rods of the exoskeleton can be simplified and the inertia of the mechanism can be reduced. The encoders installed at the joint and the self-developed human-exoskeleton interaction force sensors are used to measure the joint rotation angle and the interaction force between the wearer and the exoskeleton respectively. The human-exoskeleton interaction force sensor is designed based on the micro force sensor JLBS-MD. The micro controller STM32F407ZET6 is used to send the drive signal of the motor and receive and process the sensor feedback. In addition, a human-machine interaction interface is built based on C++ Builder, and WiFi communication is used with the micro controller, which has the functions of real-time monitoring of joint motion, data processing and command sending.

The joint movement accuracy is an important indicator for evaluating the passive training of the rehabilitation exoskeleton. However, the transmission characteristics of the compound tendon-sheath artificial muscle are severely nonlinear, and the exoskeleton has four active joints and complex structure. Therefore, the precise motion control of the exoskeleton system is difficult. To address this, we established a dynamic model of the exoskeleton system based on the transmission characteristic model of the compound tendon-sheath artificial muscle, and proposed a sliding mode control method for the motion control of the exoskeleton system.

(1) **Hip joint design**

As shown in Fig. 10.5, the left and right hip joint structure of the lower limb rehabilitation exoskeleton hip joint module is symmetrical in the sagittal plane. The flexion and extension movements are driven by the tendon-sheath artificial muscle through the hip motor assembly. An angle sensor measures the joint's rotation, providing position feedback.

As shown in Fig. 10.6a, the passive and flexible outer swing/adduction assembly of the hip joint consists of several components. These include the leaf spring fixing block, outer swing/adduction leaf spring, bearing with seat, and rotation shaft. This leaf spring mechanism is designed to prevent over-deflection, especially if patients are not fully aware, ensuring effective rehabilitation training. Additionally, specific design limits enhance patient safety by addressing potential motion control issues.

In Fig. 10.6b, a design modification replaces the leaf spring with a locking block for early-stage rehabilitation. This locking block restricts certain hip joint movements, particularly when patients may not be fully conscious of their actions. By implementing this restriction, the rehabilitation assistance becomes more effective. Moreover, keeping the tendon-sheath drive system of the hip joint static ensures optimal support. Given the complexities of the tendon-sheath artificial muscle drive, this restricted freedom will be a temporary feature of the exoskeleton setup.

The design of the hip joint leaf spring, on the one hand, increases the flexibility of the hip joint when swinging out/in from the balance position, and prevents unconscious operation to a certain extent, and releases energy to achieve passive

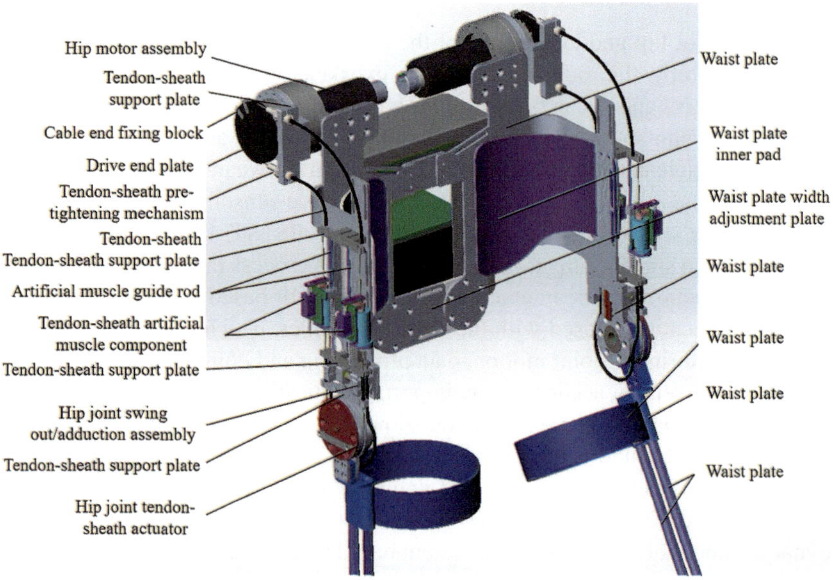

Fig. 10.5 3D model of the hip module of the lower limb rehabilitation exoskeleton [14]

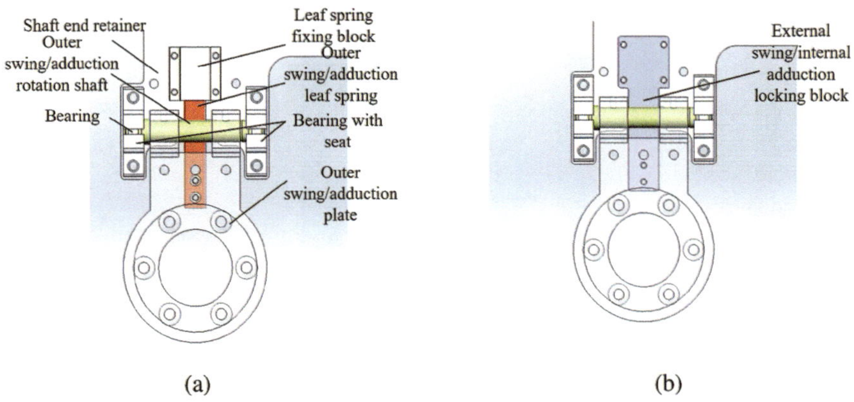

Fig. 10.6 3D drawing of the hip out-swing/in-traction assembly [14]

compliance when restoring the balance position. On the other hand, limit the range of motion of swing/retract degrees of freedom.

(2) **Knee joint design**

The knee joint of the lower limb rehabilitation exoskeleton is shown in Fig. 10.7. The compound tendon-sheath artificial muscle transmission system is employed for joint rotation. This transmission system integrates the tendon-sheath mechanism with controllable clamper and series-parallel springs.

Figure 10.8a is the structure diagram of the collet mechanism, which mainly consists of a clamping device and an electromagnet. The functional principle of the

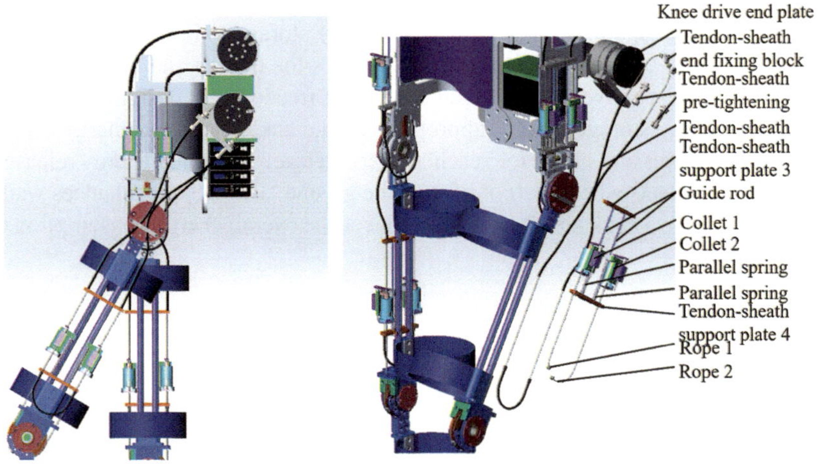

Fig. 10.7 3D diagram and explosion diagram of tendon-sheath artificial muscle transmission component [14]

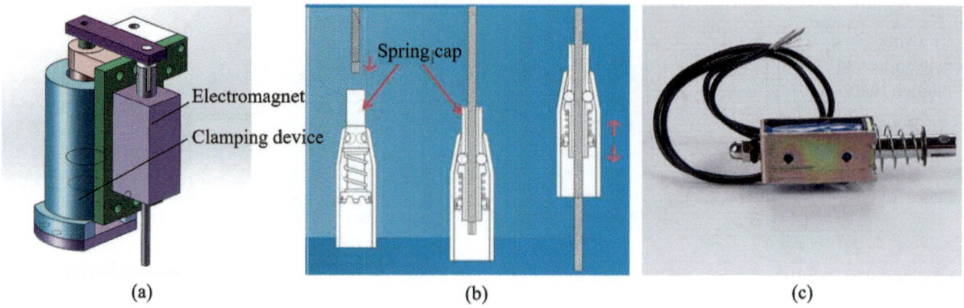

Fig. 10.8 Clutch mechanism [14]

clamping device is shown in Fig. 10.8b. The device that uses a ball roller and spring structure developed by Arakawa Giken Ltd. to clamp the rope. This device securely locks a steel wire rope upon insertion and allows adjustment or release when the top spring cap is pressed. The pressing of the spring cap is controlled by the electromagnet shown in Fig. 10.8c. This through-type push-pull DC electromagnet operates at a 24 V rated voltage, offers a 10mm stroke, and exerts a 5 N force. Together with the clamping device, this electromagnet ensures precise control over locking and releasing of the wire rope. The compound tendon-sheath artificial muscle is adopted for knee and hip joints. When collet 1 and collet 2 engage, the motor drives the corresponding joint to rotate with one spring relaxing and the other compressing. While releasing both collets gives the joint more freedom, in this case, the exoskeleton joint is disengage from the drive motor, such that the wearer can move freely.

(3) **Ankle and foot mechanism design**

The ankle joint of the lower limb exoskeleton is shown in Fig. 10.9. It comprises several key components, including the ankle joint lower leg connecting plate, toe flexion energy storage unit, back extension energy storage unit, ankle joint movement component, and foot mechanism. The ankle's toe flexion and back extension movements act as passive, under-driven degrees of freedom. By designing the stiffness of the energy storage springs appropriately, the ankle can store energy during walking. When pressure on the toe reaches a certain level, the stored energy releases, and the system resets when the foot lifts off the ground. This design enhances walking comfort, offers shock absorption, and reduces the overall energy consumption of the exoskeleton.

10.2.3.3 The Dynamic Model of the Lower Limb Exoskeleton

The lower extremity exoskeleton mechanism can be simplified as a five-bar linkage mechanism, on this basis, the Lagrangian dynamics model can be established as

$$\tau = M(\theta)\ddot{\theta} + V(\theta, \dot{\theta}) + G(\theta) \tag{10.2.1}$$

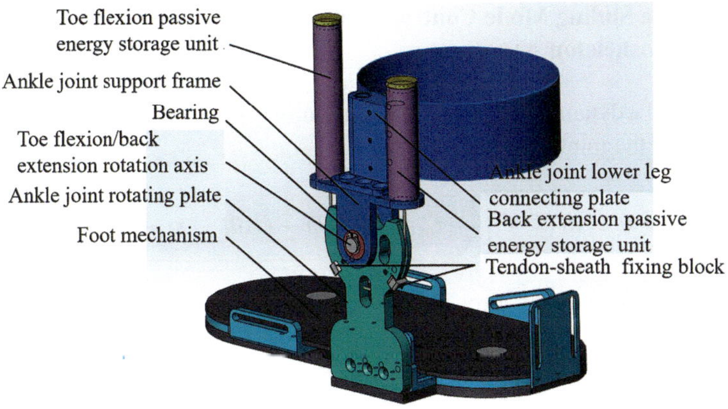

Fig. 10.9 3D modeling drawing of ankle joint [14]

in which $M(\theta)$ is the inertia matrix, $V(\theta, \dot{\theta})$ is the coriolis force vector, $G(\theta)$ is the gravity vector, $\theta, \dot{\theta}, \ddot{\theta}, \tau$ represent the angle, angular velocity, angular acceleration and joint torque of each joint.

Considering the human-machine interaction force between the lower limb exoskeleton and the wearer, the actual drive torque of the exoskeleton joint can be calculated as

$$\tau_{\mathrm{out}}(\theta, \dot{\theta}, \ddot{\theta}) = \tau(\theta, \dot{\theta}, \ddot{\theta}) + \tau_{\mathrm{fj}} + \tau_{\mathrm{h}} \tag{10.2.2}$$

in which τ_{fj} is the friction torque generated by mechanism installation, τ_h is the human-exoskeleton interaction torque. The friction torque τ_{fj} can be modeled by the Coulomb-viscous model, as Eq. (10.2.3) shows.

$$\tau_{\mathrm{fj}} = b\,\mathrm{sgn}(\dot{\theta}) + c\dot{\theta} \tag{10.2.3}$$

The joints of the lower limb exoskeleton are driven by motors through the compound tendon-sheath artificial muscle. According to the compound tendon-sheath artificial muscle transmission model, the relationship between the motor torque τ_m and the joint torque τ_{out} can be expressed as

$$\tau_{\mathrm{m}} = \frac{r_1}{r_2} \cdot \frac{\tau_{\mathrm{out}} + K_{\mathrm{PE}}\theta r_2^2 - 2\lambda F_0 r_2 \varepsilon_{\mathrm{t2}}}{\varepsilon_{\mathrm{t1}}} \quad \lambda \neq 0 \tag{10.2.4}$$

Let $\tau_{\mathrm{f}} = K_{\mathrm{PE}}\theta r_2^2 - 2\lambda F_0 r_2 \varepsilon_{\mathrm{t2}} + \tau_{\mathrm{fj}}$, $k_{\mathrm{r}} = r_1/r_2$, we can rewrite the Eq. (10.2.4) as following

$$\tau_{\mathrm{m}} = \frac{k_{\mathrm{r}}}{\varepsilon_{\mathrm{t1}}}\left[M(\theta)\ddot{\theta} + V(\theta, \dot{\theta}) + G(\theta) + \tau_{\mathrm{f}} + \tau_{\mathrm{h}}\right] \tag{10.2.5}$$

10.2.3.4 The Sliding Mode Control Method for the Lower Limb Exoskeleton

According to the dynamic model of the lower limb exoskeleton system in Eq. (10.2.5), we can obtain the joint acceleration.

$$\ddot{\theta} = M(\theta)^{-1}\left[\frac{\varepsilon_{t1}}{k_r}\tau_m - V(\theta,\dot{\theta}) - G(\theta) - \tau_f - \tau_h\right] \tag{10.2.6}$$

The joint angle tracking error is defined as

$$e(t) = \theta_d(t) - \theta(t) \tag{10.2.7}$$

where θ_d represents the desired joint motion trajectory, and $\theta(t)$ is the actual joint motion trajectory.

The sliding mode switching surface is designed as follows

$$s(t) = Ke(t) + \dot{e}(t) \tag{10.2.8}$$

in which $s(t) = [s_1 s_2 s_3 s_4]$ is the sliding mode variable vector, $K = diag[K_1 K_2 K_3 K_4]$ indicates the coefficient matrix of the proportional term of the switching function, and all the element are positive. Taking the derivation of Eq. (10.2.8), we can get

$$\dot{s}(t) = K\dot{e}(t) + \ddot{e}(t) \tag{10.2.9}$$

Substituting equations Eqs. (10.2.6) and (10.2.7) into (10.2.9), we can get

$$\begin{aligned}\dot{s}(t) &= K\dot{e}(t) + \ddot{\theta}_d - \ddot{\theta}\\ &= K\dot{e}(t) + \ddot{\theta}_d\\ &\quad - M(\theta)^{-1}\left[\frac{\varepsilon_t}{k_r}\tau_m - V(\theta,\dot{\theta}) - G(\theta) - \tau_f - \tau_h\right]\end{aligned} \tag{10.2.10}$$

The sliding mode control law is designed as follows

$$u = \tau_m + \varepsilon M(\theta)\,\mathrm{sgn}(s) \tag{10.2.11}$$

where τ_m is the equivalent control law, $\varepsilon M(\theta)\,\mathrm{sgn}(s)$ is the switching control law. ε is the switching gain matrix and all the elements are positive constants.

The equivalent control law is the approximate determinate part of the system, and needs to satisfy $\dot{s}(t) = 0$, according to Eq. (10.2.10) we can get

$$\tau_m = \frac{k_r}{\varepsilon_{t1}}[M(\theta)(K\dot{e}(t) + \ddot{\theta}_d) + V(\theta,\dot{\theta}) + G(\theta) + \tau_f + \tau_h] \tag{10.2.12}$$

In order to avoid the system chattering caused by the discontinuous sign function when the sliding mode surface is switched, we use a continuous saturation function to replace the symbolic function in Eq. (10.2.11). Such that the sliding mode controller is finally designed as shown in Eq. (10.2.13), and the sliding mode control block diagram is shown in Fig. 10.10.

$$u = \frac{k_{\mathrm{r}}}{\varepsilon_{\mathrm{t1}}}[M(\theta)(K\dot{e}(t) + \ddot{\theta}_{\mathrm{d}}) + V(\theta, \theta) + G(\theta) + \tau_{\mathrm{f}} + \tau_{\mathrm{h}}] + \varepsilon M(\theta)\,\mathrm{sat}(s) \tag{10.2.13}$$

The experiment of the lower limb exoskeleton

To verify the control accuracy of the sliding mode control method, the passive training rehabilitation experiment of the lower limb exoskeleton was carried out. In the experiment, the lower limb exoskeleton tracks the ideal joint motion curve of human walking that were measured by self-developed lightweight exoskeleton. Figure 10.11a shows the prototype of the lower limb exoskeleton driven by the compound tendon-sheath artificial muscle, and Fig. 10.11b shows the view of the passive training rehabilitation experiment.

The knee and hip motion control results of the passive training rehabilitation experiment are shown in Fig. 10.12. Figure 10.12a shows the comparison results between the ideal trajectory and the actual trajectory of the knee joint, respectively, and Fig. 10.12b shows the tracking errors of the knee joint. Figure 10.12c and d are the experimental results and the tracking errors of the hip joint. Note that θ_d represents the desired trajectory, and θ is the actual joint trajectory.

Observing the experimental results, the knee joint tracking is better than the results of the hip joint. The motion tracking of the hip joint has a lag, and its error fluctuation is relatively large. In addition, the control performance of the hip joint is slightly insufficient when the joint begin turning. This may be caused by the external swing/adduction action of the hip joint, especially when the hip joint is extended backward, the external swing is more obvious, such that it brings impact on the tracking effect. In general, with the sliding mode control method, the exoskeleton

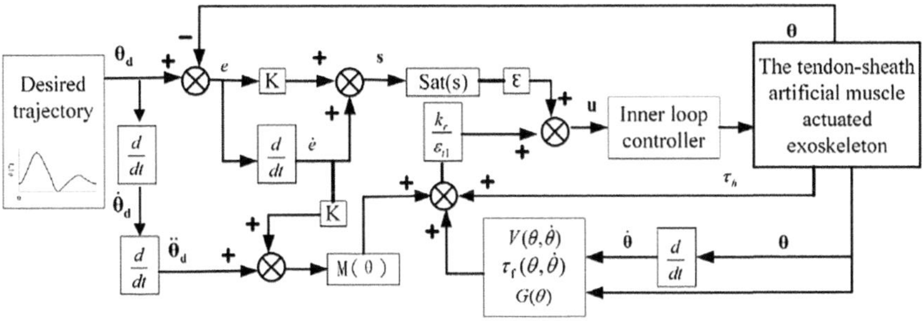

Fig. 10.10 The block diagram of the sliding mode control method for the lower limb exoskeleton [20]

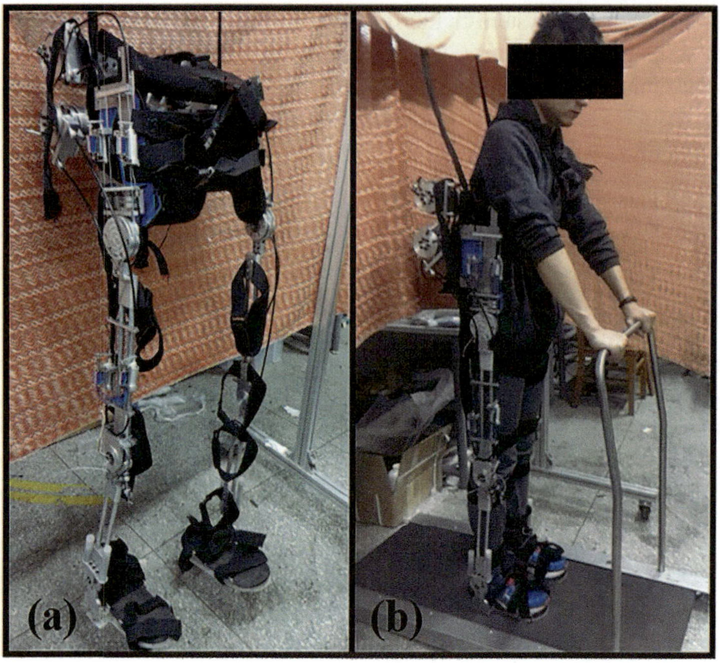

Fig. 10.11 Passive rehabilitation training experiment of the lower limb exoskeleton [20]

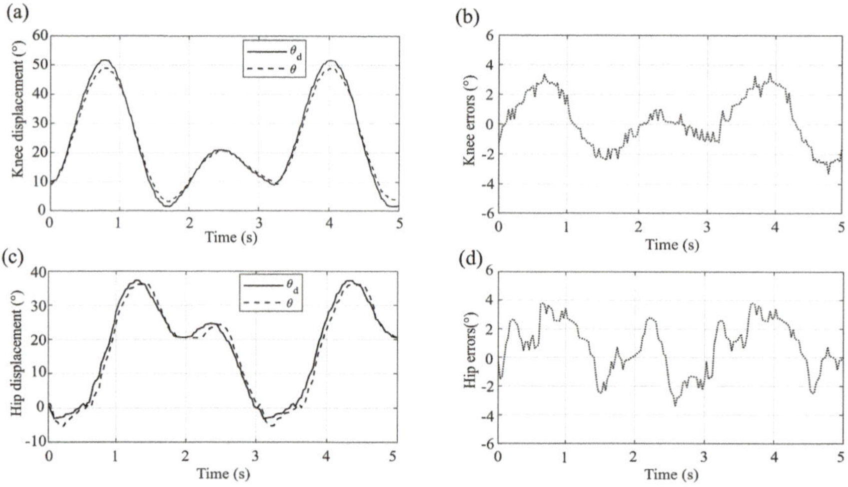

Fig. 10.12 Passive training results of the right knee joint and hip joint of the lower limb exoskeleton [20]

driven by tendon-sheath artificial muscles can achieve high precision motion, and the maximum joint tracking error is less than 3.78°, which is acceptable for the exoskeleton rehabilitation training.

10.3 Tendon-Sheath Transmitted Upper Limb Exoskeleton

The upper limb exoskeleton serves as a specialized bionic robot designed to mimic the structure of the human upper limb. Its primary function is to aid patients with movement disorders to complete rehabilitation treatments. Patients can wear the device on their affected limb, enabling the exoskeleton joints to synchronize with their natural joint movements. This chapter explores the anatomy and movement capabilities of the human arm to guide the design of the exoskeleton [17]. It then details the key components like the shoulder, elbow, wrist modules, and the tendon-sheath system.

10.3.1 Analysis of Human Upper Limbs

Based on human anatomy theory, the body locomotor system comprises bones, joints, and muscles. As Fig. 10.13 shows, bones and joints provide structural support, while skeletal muscles, innervated and controlled by the nervous system, contract to move these structures around the joints. An upper limb exoskeleton robot attaches externally to a person's arm, facilitating rehabilitation by synchronizing its movements with the individual's natural joint actions. Thus, understanding the anatomy of the human upper limb is crucial when designing such an exoskeleton.

The design of the upper limb rehabilitation exoskeleton follow the bionics principles, emphasizing optimal man-machine coordination. It's vital to align rotation center of each motor joint with the patient's upper limb during rehabilitation. Any misalignment between the patient's limb and the exoskeleton can cause undesired pulling during exercises. This misalignment can restrict movement, hinder effective training, and potentially harm the patient.

Traditionally, upper limb exoskeletons simplify the human arm (excluding hand joints) into a seven-degree-of-freedom (DOFs) structure. This structure consists of specific joint movements, as illustrated in (Fig. 10.14). These include three DOFs for shoulder movements, two for the elbow, and two for the wrist. To ensure safety, the joint movements of the exoskeleton should stay within the natural range of the human body. Table 10.1 shows the ultimate motion range of each joint of the human body and the motion range in daily life.

In the simplified human upper limb model, shoulder joint movement is often reduced to the motion of the glenohumeral joint alone. However, as shown in Fig. 10.15, both the sternoclavicular joint (point O) and acromioclavicular joint (point C) contribute to shoulder elevation/depression and forward/backward swings during

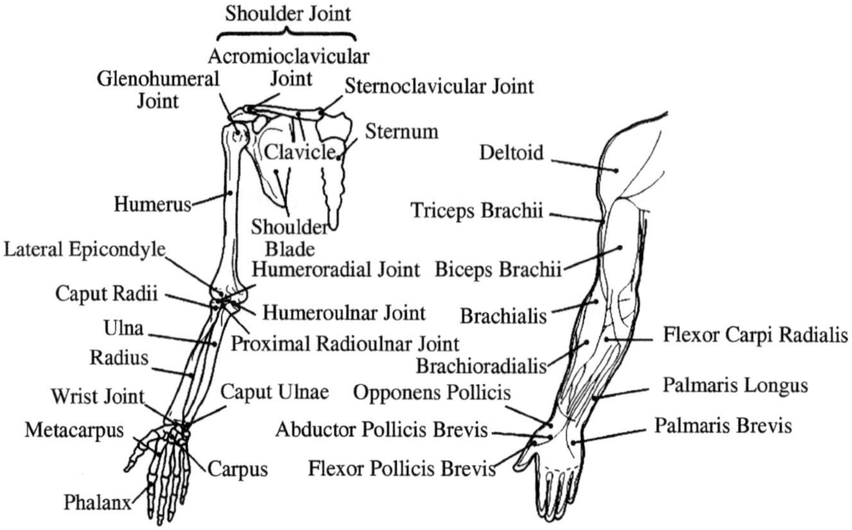

Fig. 10.13 The schematic diagram of the main skeleton and muscle of human upper limb [17]

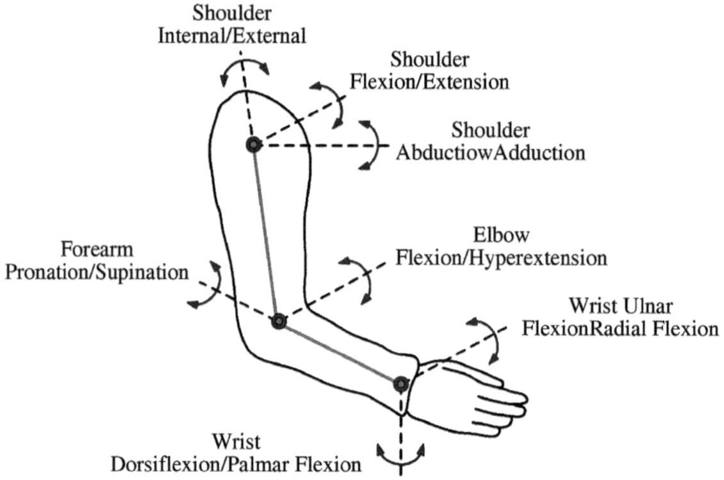

Fig. 10.14 The schematic diagram of a simplified model of human upper limb motion [17]

upper limb movements. This results in a shift in the rotation center of the glenohumeral joint (point S). To ensure proper coordination between the exoskeleton and the human upper limb, the design of the exoskeleton must account for these deviations in the human shoulder joint, preventing misalignment of the joint rotation center. Based on human anatomy and image processing technology, Leonard Freedman et al. analyzed the characteristics of changes in the position of shoulder joints during the movement of human upper limbs. It was found that the glenohumeral joint

Table 10.1 Range of upper limb joint motion

Joint	Joint motion	Limit range of motion (°)	ADL range of motion (°)
Shoulder	Flexion	155	135
	Extension	−60	−45
	Abduction	150	135
	Adduction	−35	−25
	External	80	65
	Internal	−60	−50
Elbow	Flexion	55	45
	Hyperextension	−105	−90
	Pronation	95	90
	Supination	−85	−75
Wrist	Ulnar flexion	35	25
	Radial flexion	−20	−10
	Dorsiflexion	35	30
	Palmar flexion	−65	−55

changed its position in both vertical and horizontal directions during external pendulum/adduction. Based on the CT detection data of human shoulder joint, Tobias Nef et al. fitted the rotation center position of glenohumeral joint as a function of the external pendulum/adjunct angle, as shown in Eqs. (10.3.1) and (10.3.2).

$$x_s = \left[-|OC|_{\text{ref}} \cdot \cos(\alpha - 90^\circ) + |SC|_{\text{ref}} \cdot \sin(\alpha + \beta)\right] \cdot \frac{H}{H_{ref}} \tag{10.3.1}$$

$$y_s = \left[|OC|_{\text{ref}} \cdot \sin(\alpha - 90^\circ) + |SC|_{\text{ref}} \cdot \cos(\alpha + \beta)\right] \cdot \frac{H}{H_{\text{ref}}} \tag{10.3.2}$$

where,

$$\alpha = \begin{cases} 15\left[10\left(\frac{\theta}{80^\circ}\right)^3 - 15\left(\frac{\theta}{80^\circ}\right)^4 + 6\left(\frac{\theta}{80^\circ}\right)^5\right] + 99^\circ & 0^\circ \le \theta < 80^\circ \\ 114^\circ & 80^\circ \le \theta < 140^\circ \\ 9\left[10\left(\frac{\theta-140^\circ}{40^\circ}\right)^3 - 15\left(\frac{\theta-140^\circ}{40^\circ}\right)^4 + 6\left(\frac{\theta-140^\circ}{40^\circ}\right)^5\right] + 114^\circ & 140^\circ \le \theta < 180^\circ \end{cases}$$

$$\beta = \begin{cases} 143^\circ & 0^\circ \le \theta < 70^\circ \\ 35\left[10\left(\frac{\theta-70^\circ}{70^\circ}\right)^3 - 15\left(\frac{\theta-70^\circ}{70^\circ}\right)^4 + 6\left(\frac{\theta-70^\circ}{70^\circ}\right)^5\right] + 143^\circ & 70^\circ \le \theta < 150^\circ \\ 178^\circ & 150^\circ \le \theta \le 180^\circ \end{cases} \tag{10.3.3}$$

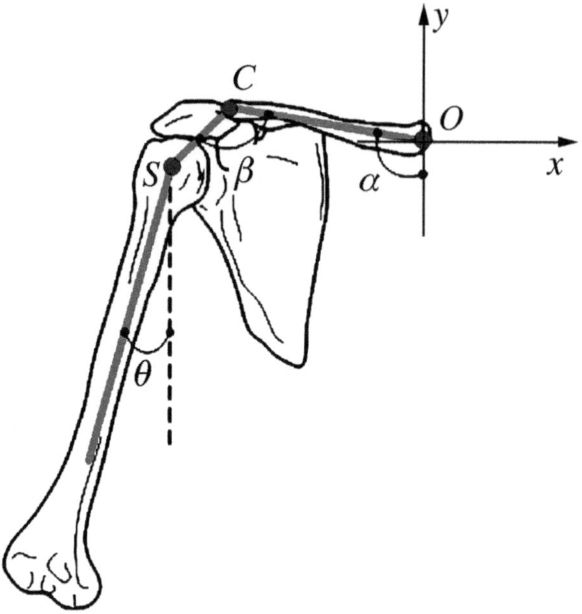

Fig. 10.15 The schematic diagram of shoulder joint angle [17]

where H and H_{ref} represent the actual height and reference height of the human body, respectively; $|OC|_{ref}$ and $|SC|_{ref}$ represents the reference joint spacing; α, β, and θ represent the joint angles of the sternoclavicular, acromioclavicular, and glenohumeral joints, respectively; x_s and y_s represent the amount of position change in the horizontal and vertical directions of the glenohumeral joint when the glenohumeral joint changes in the range of [0°, 180°] (corresponding to the human height of 1.7 m), as shown in Fig. 10.16. According to the CT test data, when *Href* is 1.8 m, the $|OC|_{ref}$ is equal to 149.4 mm, $|SC|_{ref}$ equals 66.8 mm.

10.3.2 Human-Machine System Structure Design Requirement

Given the unique characteristics of the upper limb rehabilitation exoskeleton system, its structural design must fulfill specific requirements:

1. Adaptability to various rehabilitation needs:
   The exoskeleton must cater to the diverse training needs of various rehabilitation subjects. Given the variations in height, weight, and shape of patients' injured limbs, the exoskeleton's dimensions should be adjustable accordingly. Notably,

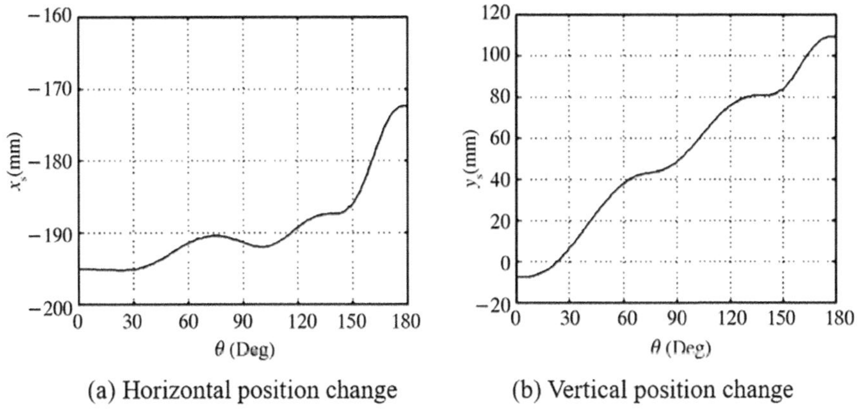

Fig. 10.16 The schematic diagram of rotation center of glenohumeral joint [17]

patients with hemiplegia may experience limb swelling due to prolonged inactivity. Therefore, the human-machine interface of the exoskeleton should offer a broad adjustment range to accommodate varying sizes while ensuring comfort.

2. Safety in rehabilitation training:
   Safety remains paramount. The exoskeleton must not harm the injured limb. Its joint motion range should align with the capabilities of the injured limb. Considering that patients may lack the reflexes to protect themselves during emergencies, an emergency stop switch is essential. This switch allows patients to halt the system using their unaffected hand or foot, ensuring no harm even during power outages or system malfunctions. Crucially, the exoskeleton must align with the joint rotation centers, especially around the shoulder joint.
3. Optimal lightweight design:
   The exoskeleton structure should prioritize being lightweight and compact. This design minimizes the weight and size of the moving parts, conserves energy, and maintains adequate strength and stiffness for effective rehabilitation training.
4. Versatility in training intensity:
   Rehabilitation needs can vary widely among hemiplegia patients in terms of intensity and mode. Hence, the exoskeleton design should accommodate different motion speeds, ranges, forces, and training modes. Importantly, the interaction force between the exoskeleton and the patient must remain below safety thresholds, ensuring effective and safe rehabilitation.

10.3.3 Structure Design of Tendon-Sheath Transmitted Upper Limb Rehabilitation Exoskeleton

Figure 10.17 displays the structure of the upper limb rehabilitation exoskeleton, it includes components like the exoskeleton actuator, adaptive mobile platform, tendon-sheath transmission module, mobile bracket, adjustable chair, industrial control computer, and display module. The actuator integrates three sections: shoulder, elbow, and wrist joint motion modules. The adaptive platform ensures the shoulder joint of the exoskeleton aligns with human movement variations. Meanwhile, the tendon-sheath module, affixed to a mobile bracket, drives the exoskeleton shoulder and elbow motions. Individuals undergoing rehabilitation utilize a height-adjustable chair for training. The industrial control computer, equipped with data acquisition, analog output, and motion control cards, manages the training strategies and provides real-time feedback. The main components are described in detail below.

(1) **Tendon-sheath transmission shoulder joint motion module**

Figure 10.18 shows the shoulder joint of the upper limb rehabilitation exoskeleton, which includes three primary motion joints: internal/external rotation, abduction/adduction, and flexion/extension. The three degrees of freedom (3 DOFs) are driven through band transmission mechanism. Among them, J1, J2 and J3 are rotation axis of internal/external rotation, abduction/adduction, and anterior flexion/extension of shoulder joint respectively, and they intersect at point S. Before rehabilitation training, the physician can adjust the height of chair to make point S coincide with the glenohumeral joint of the patient. The shoulder joint adopts tendon-sheath transmission to drive each joint. Each tendon-sheath transmission module has two sheathes and wire ropes to control one DOF. One end of the sheath is fixed to the pretension baffle of the tendon-sheath transmission module, and the other end is fixed to each joint's fixed bracket of sheath. One end of the wire rope is connected with the drive pulley of the tendon-sheath transmission module, and the other end is connected with the load pulley of the corresponding joint through the sheath. The sheath ends affix to tension baffles and joint brackets, while wire rope ends link drive and load pulleys. This setup effectively separates the exoskeleton actuator and drive unit, reducing weight and size while meeting design criteria for low power and lightweight.

The shoulder joint adopts rotary potentiometer to detect the joint position. The back of the shoulder joint is installed with a stretching spring K1, and the moving end of this spring is connected with the shoulder joint wire rope. A stretching spring (K1) aids in transferring forces between the shoulder and elbow joint modules, facilitated by a self-regulating pulley module ensuring pulley and wire alignment.

(2) **Tendon-sheath transmission elbow joint and wrist joint motion module**

Figure 10.19 shows the elbow joint and wrist joint motion modules of the exoskeleton. The elbow module handles bending and forearm rotation, while the wrist module manages wrist movements. Before training, doctors adjust these modules to fit the patient's joints precisely. Tendon-sheath mechanisms drive elbow actions,

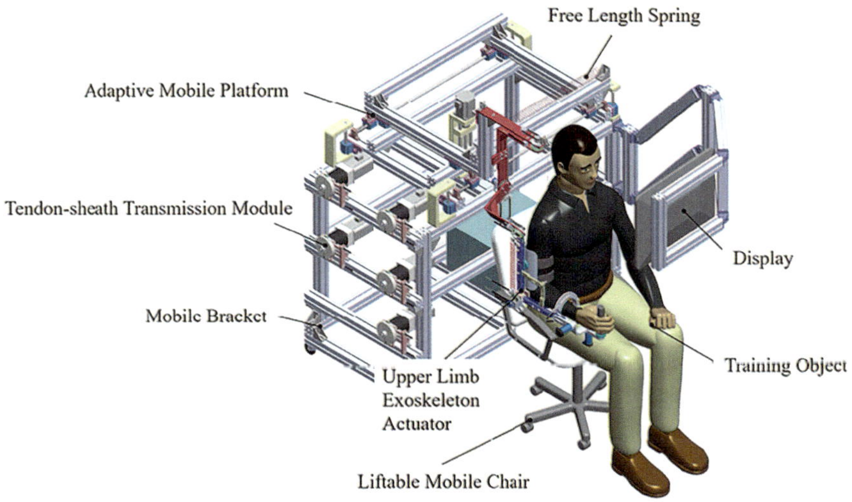

Fig. 10.17 The schematic diagram of the overall structure of upper limb rehabilitation exoskeleton [17]

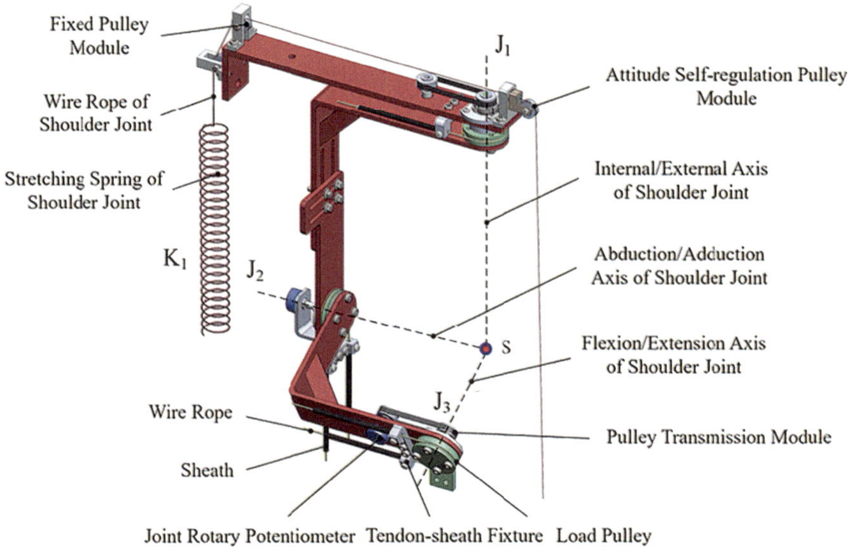

Fig. 10.18 The schematic diagram of upper limb rehabilitation exoskeleton shoulder joint motion module [17]

with potentiometers measuring angles. A spring (K2) supports the upper arm of the exoskeleton. The wire rope of the elbow connects via a flexible chute. The adjustable parallel rods enhance the elbow motion to form a stable structure. A secure strap and

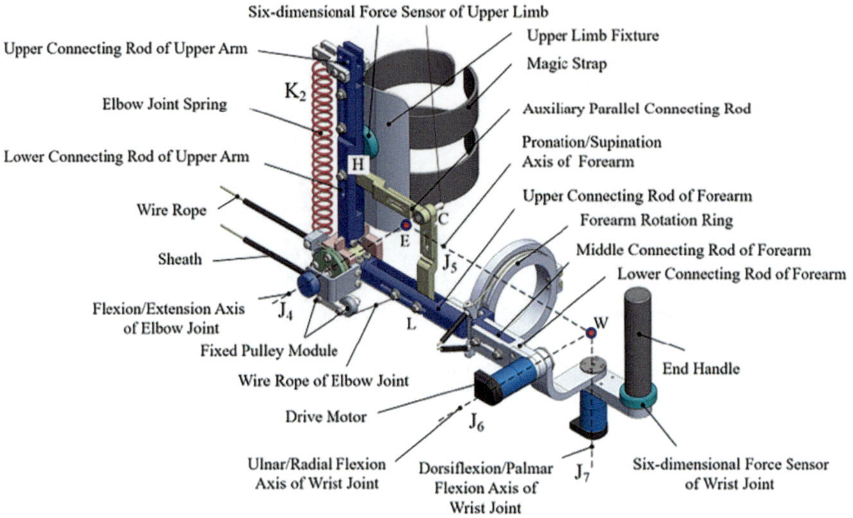

Fig. 10.19 The schematic diagram of upper limb rehabilitation exoskeleton elbow and wrist joint motion module [17]

force sensor ensure the patient's arm fits comfortably and safely. Patients use an end handle, also equipped with a force sensor, for rehabilitation exercises.

10.3.3.1 Shoulder Joint Adaptive Mobile Platform

(3) **Shoulder joint adaptive mobile platform**

Figure 10.20 displays the adaptive mobile platform designed to align the exoskeleton shoulder joint with the human shoulder joint during movement. This platform adjusts to ensure both joints align, preventing misalignment and potential injury. It features linear guide rails for movement in three directions: horizontal (X and Y) and vertical (Z). Springs in the horizontal rails help maintain initial positions but adjust if the human shoulder shifts. A ball screw, driven by a servo motor, controls vertical movement. The height of the moving platform is determined according to the angle of shoulder abduction/adduction. According to the built-in encoder and screw, the motor angle corresponding to the height of the platform can be obtained, and the real-time adjustment of the height of the platform can be finally realized. This platform ensures synchronized and safe movement between the human and exoskeleton shoulders.

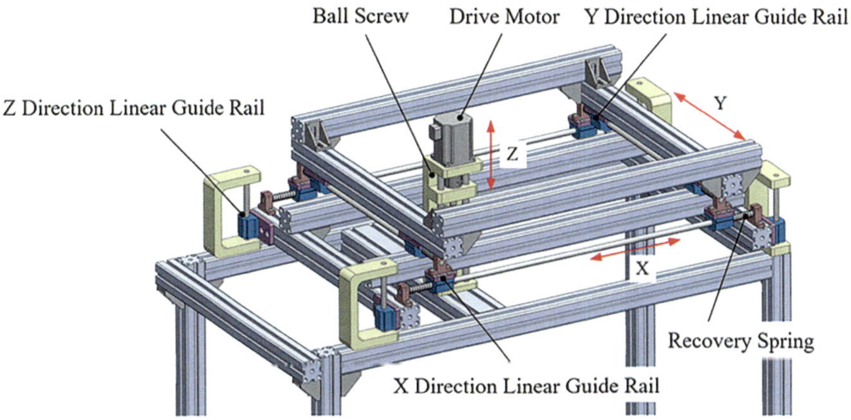

Fig. 10.20 The schematic diagram of upper limb rehabilitation exoskeleton shoulder joint adaptive mobile platform [17]

10.4 Tendon-Sheath Transmitted Passive Light Exoskeleton

Active exoskeletons can handle different walking conditions, but require bulky components (e.g., motors) that need a significant source of power to do so. Purely passive exoskeletons are lightweight and energy-neutral, containing energy-recycling mechanisms that capture energy loss during negative power phases and return it as walking assistance. However, they are usually designed for stereotyped gaits (e.g., walking at fixed speed) and thus show poor adaptivity for variable conditions. To overcome these issues, we present a novel design for a quasi-passive ankle exoskeleton that recycles the heel-strike energy loss at early stance for propulsion at late stance [18, 19]. This design integrates the merits of both active and passive exoskeletons, and captures the heel-strike energy loss and recycles it into propulsion.

10.4.1 Design of Light Ankle Exoskeleton

The key aspects of our design include (i) an elastic energy-storage mechanism at heel, and (ii) an electrically controlled clutch mounted at the back of shank frame (Fig. 10.21). The clutch is triggered by a pressure signal from the film sensor (not shown) fixed on the forefoot plate with the battery and control panel off-board. These two elements of the design (detailed below) are then coupled together by a stiff extension spring, termed assistance spring. One end of the assistance spring connects to a wire rope, which is tied to a reel. A torsion spring inside the reel prevents the rope from going slack by wrapping it onto reel surface. The other end of the assistance spring connects to a Kevlar rope, which is fixed on the back beam of the elastic energy-storage mechanism. The torsion spring force is about 1.2 N,

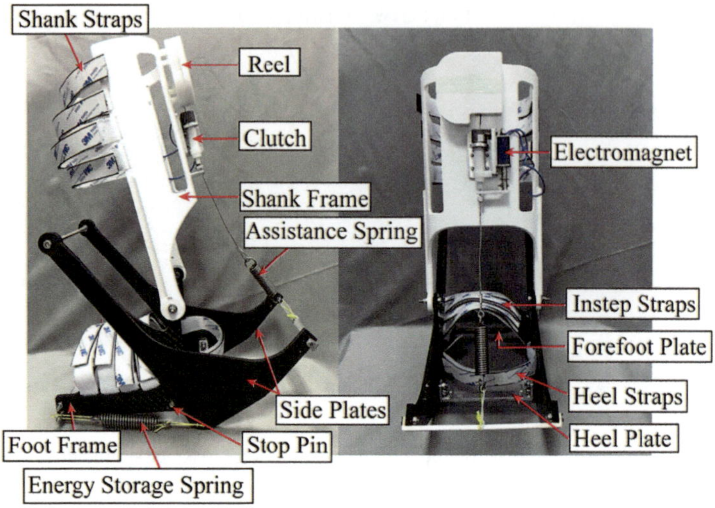

Fig. 10.21 The mechanical design of the ankle exoskeleton. The heelstrike energy-storage mechanism consists of two springs and two side plates, one on either side of the foot (Reprinted from [19] with permission from IEEE)

which exerts an almost negligible resistance (0.14 Nm) on ankle motion. The total mass of our prototype is 1.67 kg (754 g per leg).

The exoskeleton is designed to capture the collision energy of heel-strike and return it for push-off. When foot makes initial contact with ground (heel-strike), the clutch is in disengagement state, allowing free ankle rotation. Heel-strike occurs from the moment when the lower edges of side plates makes contact with ground (Fig. 10.22a) to the moment when the heel is stopped by ground (Fig. 10.22b). Thus, the negative work during heel-strike, or part of it, is transformed into elastic potential energy stored in the stretched springs of the elastic energy-storage mechanism. As the user plantarflexes to foot-flat, the reel above the clutch takes up the slack in the rope. When the increasing forefoot pressure signal (a voltage signal) exceeds a threshold, the clutch engages the wire rope (Fig. 10.22c).

As soon as the user begins to dorsiflex, the assistance spring stretches for storing energy. During late stance, the heel rises (Fig. 10.22d), the energy stored in the energy-storage mechanism is first transferred to the assistance spring (Fig. 10.22e), and then all the elastic energy gathered in the assistance spring is returned to the ankle, assisting plantarflexion. Finally, when the forefoot leaves ground (Fig. 10.22f), the decreasing pressure signal falls below another threshold, and the clutch disengages the wire rope to allow free rotation of the ankle, so that the user can dorsiflex their ankle naturally without interference during swing phase (Fig. 10.22g). This cycle is repeated in the following steps.

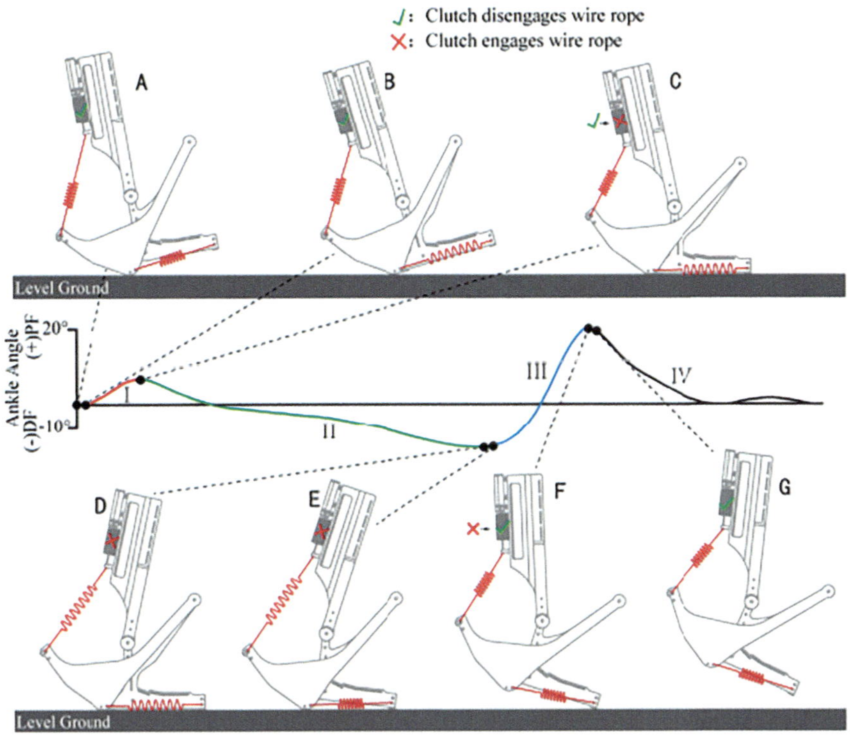

Fig. 10.22 Ankle exoskeleton prototype mechanical function (Reprinted from [19] with permission from IEEE)

The function of the arches of human foot inspired us when designing this exoskeleton. Owing to its elastic property, the thick plantar aponeurosis of the medial longitudinal arch can store the energy of ground contact reaction force, returning it at the next step and thus reducing the cost of walking and, particularly, running, where vertical forces are higher. Illumined by the structure of a foot arch, two highly rigid carbon-fiber side plates are pivoted on either side of the foot frame, forming two triangle truss-spring structures alongside the foot, one on either side (Fig. 10.23). The base of the triangle is a linear tension spring, which is used for absorbing the impact of heel-strike. The two side plates are rigidly connected by a beam, which is attached to the assistance spring by a Kevlar rope. The triangle truss-spring structure has two states: Initial State (Fig. 10.23a) and Energy Storage State (Fig. 10.23b).

In the Initial State, the resultant force, F_1, is exerted by the tension springs on the side plates, which are stopped by pins fixed in position A on either side of the foot frame. In the Energy Storage State, the bottom of foot frame and the lower edges of side plates are in contact with ground, while the two tension springs are stretched. The dynamic process from Initial State to Energy Storage State corresponds to the process from heel-strike (Fig. 10.22a) to foot-flat (Fig. 10.22c).

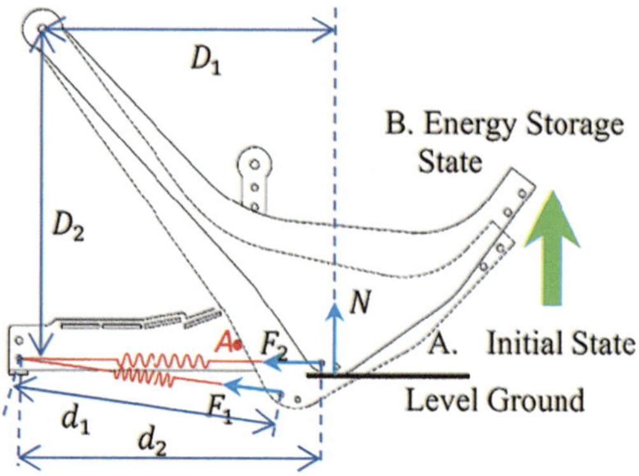

Fig. 10.23 The dynamic process from the initial state (**a**) to the energy storage state (**b**) (Reprinted from [19] with permission from IEEE)

During late stance, the ankle plantarflexes to push-off, and the assistance spring operates in parallel with the users calf muscles. The range of the coefficient of spring stiffness is jointly determined by ankle angle during stance and resultant force of the tension springs in the energy-storage mechanism. To implement plantarflexion assistance, the maximum force of assistance spring, F_3, at the maximum dorsiflexion angle (Fig. 10.24a) should satisfy Eq. (10.4.1).

$$F_3 D_3 < F_1 D_2 \tag{10.4.1}$$

where F_1 (248 N) is the resultant force of two tension springs in the Initial State as described above, and D_2 and D_3 are the vertical distances from the axis of rotation of side plates to the lines of action of F_1 and F_3. According to the measurement of joint angle during walking, the maximum dorsiflexion angle is about 12°, and D_2 and D_3 are 189 mm and 167 mm, respectively. According to Eq. (10.4.1), the value of F_3 should be less than 280 N. Since the maximum elongation of the assistance spring is 71 mm (calculated by $\Delta s = s_1 - s_2$, where s_1 is the maximum length of the assistance spring, as shown in Fig. 10.24a. s_2 is the initial length of assistance spring at foot-flat moment, as shown in Fig. 10.24b), if the initial tension of the assistance spring is 0 N at foot-flat, the spring stiffness coefficient should be less than 3.9 N/mm. We used an extension spring with a stiffness of 3.5 N/mm as the assistance spring.

10.4.2 Control Method of Exoskeleton Energy-Saving Walking

According to the description of exoskeleton function in a typical gait cycle (Fig. 10.22), it is imperative that the clutch engages and disengages the assistance spring in a timely manner to ensure effective assistance torque on ankle. Therefore, a timely and correct feedback on contacting state between the forefoot and ground is the key to determine when the electromagnet should be energized. We noticed that

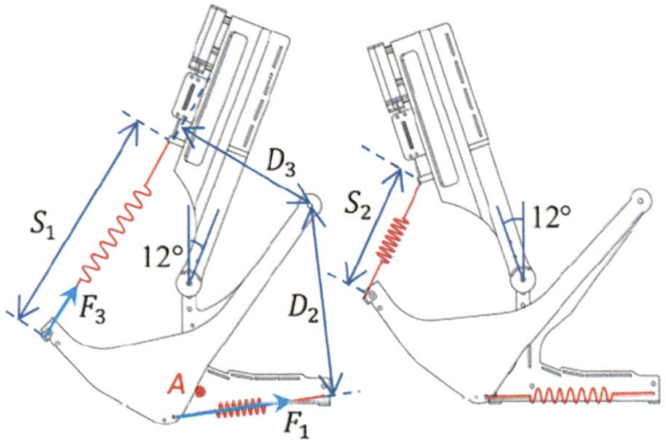

Fig. 10.24 Schematic diagrams of length change of the assistance spring (Reprinted from [19] with permission from IEEE)

the previous passive exoskeleton use ankle angle feedback to control its clutch [15]. Once the recorded ankle angles are reached, the assistance spring is engaged or disengaged. The main drawback of this approach is that the passive clutch is unable to accommodate variations in ankle angles of different users. To alleviated this issue, the pressure feedback from forefoot is used to control the engagement/disengagement of our clutch. A thin film pressure sensor (FlexiForce A301 Sensor) is fixed on the forefoot plate of the foot frame to detect gait events. When the pressure is greater or less than specified threshold values, the microcontroller (STM32 board) sends a high-level voltage acting as a input signal to control a relay (DELIXI CDG1-1DD/10A), which is used to control the driving circuit of the electromagnet. Once the circuit is turned on, the electromagnet will be energized to exert a pressing force on the upper ratchet of the switching device. For the rest time of gait cycle, the relay is off and the clutch maintains an engagement or disengagement state.

During the foot-flat stage, the voltage of forefoot pressure signal increases (Fig. 10.25). When the signal voltage exceeds a threshold of 0.02 V for the first time, to prevent accidentally triggering the clutch, the microcontroller compares the signal voltage with the threshold value after a 10 ms delay (Fig. 10.26a). If the new signal voltage exceeds the threshold value after the delay, it is concluded that the foot-flat motion is completed. Then, the microcontroller sends a high-level signal to energize the electromagnetic circuit for 15 ms, during which the electromagnet exerts a pressing force on the clutch for state switching, from disengagement to engagement.

During the toe-off stage at late stance, the voltage of forefoot pressure signal decreases (Fig. 10.25). We set the threshold to 0.01 V. When the signal voltage falls below 0.01 V for the first time, for the same reason of avoiding false triggering, the microcontroller compares the signal voltage with the threshold twice, each after a 10 ms delay (Fig. 10.26b). If the newest signal voltage is below the threshold, it is concluded that the toe-off motion is completed. The microcontroller then sends a highlevel signal to the relay after a 50 ms delay. The electromagnet circuit will be

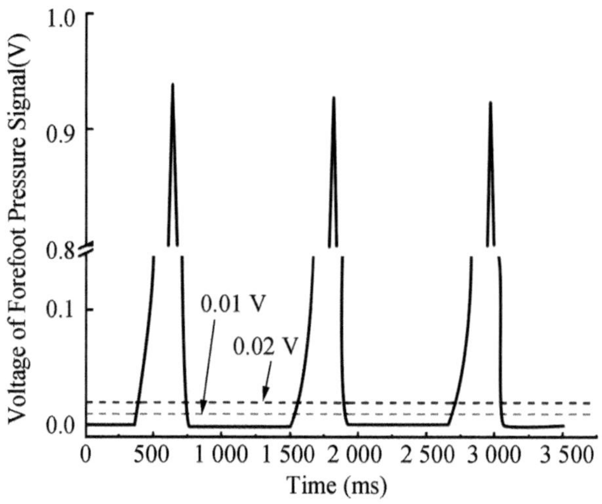

Fig. 10.25 Pressure signal at forefoot during walking at 1.25 m/s and the two threshold values (0.01 and 0.02 V) (Reprinted from [19] with permission from IEEE)

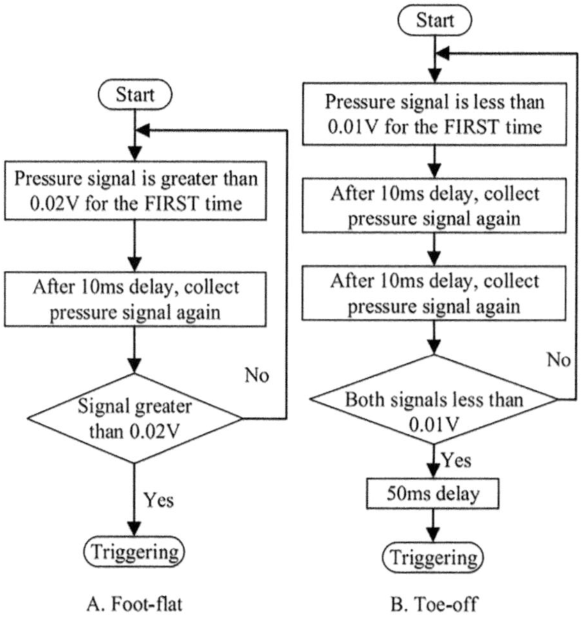

Fig. 10.26 Schematic block diagrams of the triggering process of the clutch when forefoot pressure increases (**a** Foot-flat) or decreases (**b** Toe-off) (Reprinted from [19] with permission from IEEE)

energized for 15 ms for state switch of the clutch, from engagement to disengagement. The 50 ms delay arises from the mechanical design of the clamper. When the rope is under high loading force, the static friction force between the steel balls and inner wall of upper clamper can be greater than the pressing force of the electromagnet, thus the rope may not be disengaged. Therefore, the tension of the series spring must be released before the clutch disengages the rope. We consider this mechanical character of the clutch as a built-in failsafe for a situation in which the disengagement of the clutch is triggered too early. This characteristic makes controlling the clutch a simple operation.

10.4.3 Validation Experiments and Discussion

(1) **Experiment setup**

To investigate the functionality of our ankle exoskeleton on real users, we conducted an experiment with six healthy male subjects (age: 28 ± 4 years; weight: 77 ± 8.5 kg; height: 1.79 ± 0.08 m) walking on a treadmill. The experimental setup is shown in Fig. 10.27. To evaluate the effect of recycling heel-strike energy loss on walking, each subject walked on a treadmill at a fixed speed of 1.25 m/s under three conditions: Baseline, the subject walked without the exoskeleton; No Recycle (NR), the subject walked wearing bilateral exoskeletons while the springs of the heel-strike energy-storage mechanism were removed, but the assistance springs were retained; Complete Exoskeleton (CE), the subject walked with the complete exoskeletons on both legs.

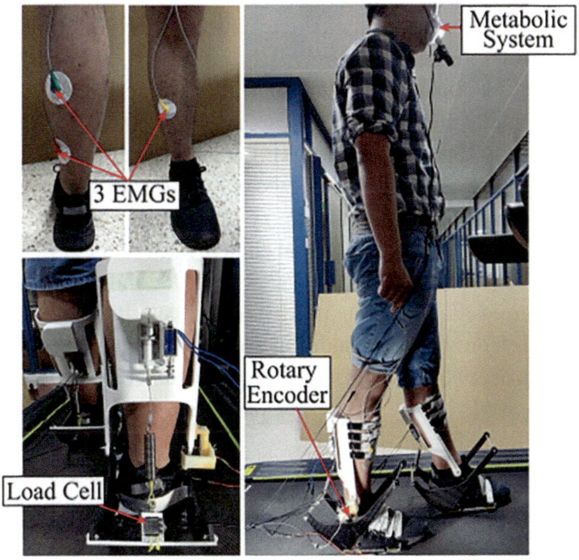

Fig. 10.27 Experimental setup for participant walking with the ankle exoskeleton on a treadmill and EMG sensor placement (Reprinted from [19] with permission from IEEE)

We defined the ankle angle as 0 at the upright standing posture and ankle plantarflexion was positive. The exoskeleton torque was calculated as the product of the smoothed spring force and the perpendicular distance from the center of rotation of ankle joint to the line of action of assistance spring. The mechanical power delivered by the exoskeleton was the product of the torque and angular velocity (derivative of ankle angle). These torque and power values were then normalized to individual body mass. Muscle activities were measured from the soleus (SOL), gastrocnemius (GAS) and tibialis anterior (TA) using a wireless electromyography (EMG) system (Sichiray, Jiangsu, China). The metabolic rate of subject was estimated by measuring the rates of O_2 consumption and CO_2 production using a portable gas-exchange measurement device (Consensic, Jiangsu, China). The rates of O_2 consumption and CO_2 production were converted to metabolic rate by applying Brockways standard equations [16]. The net metabolic rate of tasked walking was calculated by subtracting the metabolic rate during quiet standing from the metabolic rates recorded during walking. These net metabolic rate values were then normalized to each individual's body mass.

(2) **Experimental results and discussion**

Figure 10.28 (Top) shows the average ankle angles from the six subjects under the three conditions. The data were normalized to stride period. Joint angles changed little across conditions during the swing phase (60–100% stride), confirming that the disengagement of the clutch occurred in a timely manner after toe-off, and the exoskeleton did not interfere with normal ankle functions (e.g., toe clearance). Figure 10.28 (Middle) shows that due to the introduction of the heel-strike energy-storage mechanism, the peak exoskeleton torque increased by 22.7% (from 0.22 ± 0.02 Nm/kg to 0.27 ± 0.03 Nm/kg); and the peak exoskeleton power (Fig. 10.28 (Bottom)) increased by 86.8% (from 0.38 ± 0.06 W/kg to 0.71 ± 0.09 W/kg).

Figure 10.29 presents the average activities of SOL (Top), GAS (Middle) and TA (Bottom) over the whole stride under the three conditions. SOL and GAS activities were reduced under NR and CE conditions compared with Baseline during early and mid-stance 0–40% stride). TA activity seemed to increase during early and mid-stance under NR and CE conditions. To evaluate the muscle activities statistically over all subjects, Fig. 10.30 presents the averages of change rates of the EMG RMS across the six subjects on all measured muscles under NR and CE conditions compared with the Baseline. It is apparent that the reductions in EMG activities of SOL ($6.8 \pm 1.4\%$) and GAS ($6.6 \pm 1.0\%$) under CE condition are greater than that in NR($6.1 \pm 1.3\%$ and $5.8 \pm 1.0\%$, respectively) because of the recycled heel-strike energy. The EMG activity of TA increased slightly during NR and CE compared with the Baseline, by $4.1 \pm 0.8\%$ and $4.4 \pm 0.8\%$, respectively.

The net metabolic rate of each subject is reported in Fig. 10.31a. Result of the ANOVA for the factor assistance mode indicates that exoskeleton assistance affects the metabolic rate ($p < 0.0001$). The results of post-hoc paired t-test show that NR mode reduced the metabolic rate of walking to 2.97 ± 0.34 W/kg (mean $\pm$SD), down from 3.10 ± 0.33 W/kg for normal walking (Baseline), a reduction of $4.1 \pm 1.2\%$ ($p < 0.05$). CE mode reduced the metabolic rate of walking to 2.90 ± 0.33 W/kg by

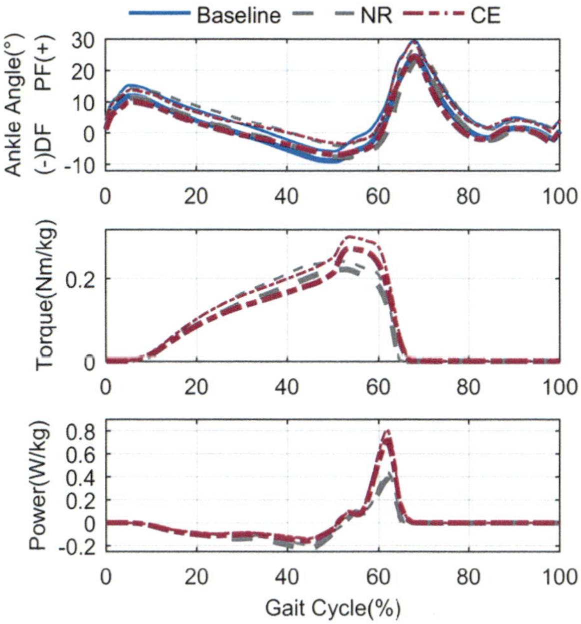

Fig. 10.28 Time courses of ankle angles (Top), calculated exoskeleton torques (Middle) and powers (Bottom) averaged across six subjects under baseline, no recycle (NR) and complete exoskeleton (CE) conditions in validation study I (Reprinted from [19] with permission from IEEE)

a reduction of $6.4 \pm 1.3\%$ ($p < 0.05$) relative to the Baseline (Fig. 10.31b). We concluded that the heel-strike energy-storage mechanism enhanced walking assistance ability of the exoskeleton.

This work demonstrates the way in which the merits of an active exoskeleton and a passive exoskeleton can be combined to create a new kind of quasi-passive device. The quasi-passive ankle exoskeleton recycles the heel-strike energy loss at early stance for propulsion at late stance. In this context, Quasi indicates that the exoskeleton consumes a negligible amount of electrical energy to drive the electronics and the clutch compared with traditional powered exoskeletons. Passive means that the exoskeleton has no actuators (e.g., motors). We aimed to combine the benefits of both kinds of exoskeletons the adaptivity of the powered and the energy economy of the passive to provide assisting ankle torque that can reduce muscle demands (i.e., force or power) and widen the application scenarios of exoskeleton.

The previous passive exoskeletons can only offer fixed mechanical properties, because the engagement/disengagement of their clutches is an explicit function of joint angle. Our prototype mitigated this limitation and expanded the range of application to situations such as incline/decline slopes. To accomplish these improvements, we developed a novel clutch driven by an electromagnet, which was controlled by the pressure signal from forefoot. We integrated a heel-strike energy-storage mechanism with a foot frame. The clutch and recycling mechanism were coupled together

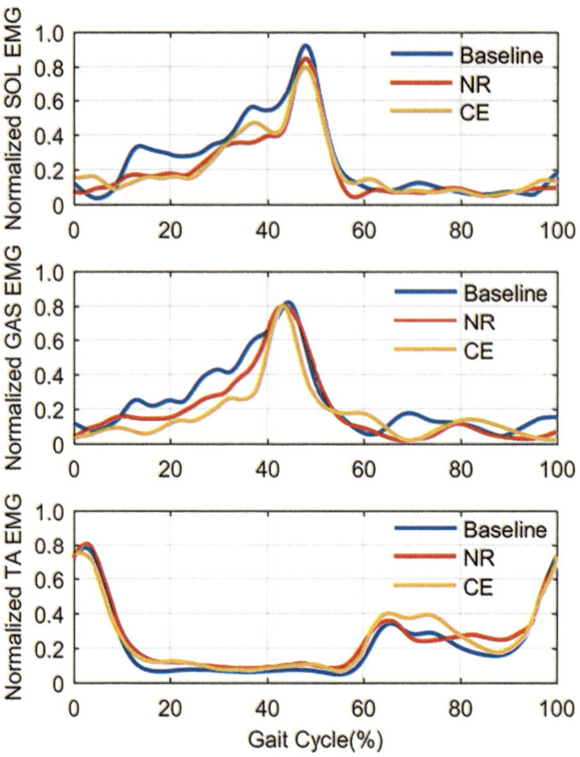

Fig. 10.29 Time courses of the filtered EMG signals of SOL (Top), GAS (Middle) and TA (Bottom) averaged across six subjects under baseline, no recycle (NR) and complete exoskeleton (CE) conditions in validation study I (Reprinted from [19] with permission from IEEE)

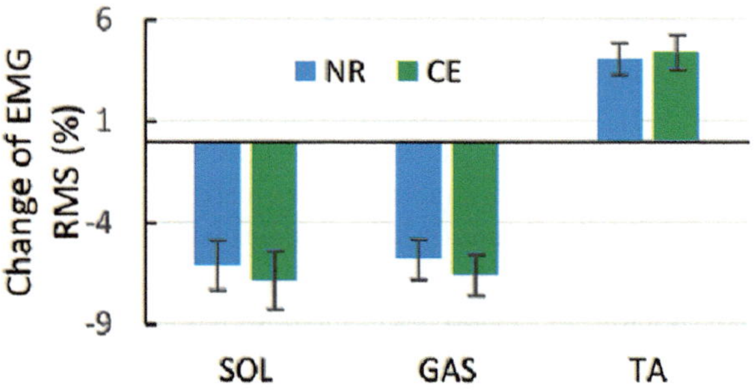

Fig. 10.30 Changes of EMG RMS (mean ±1 SD, $N = 6$) under no recycle (NR) and complete exoskeleton (CE) assistance compared with Baseline walking (Reprinted from [19] with permission from IEEE)

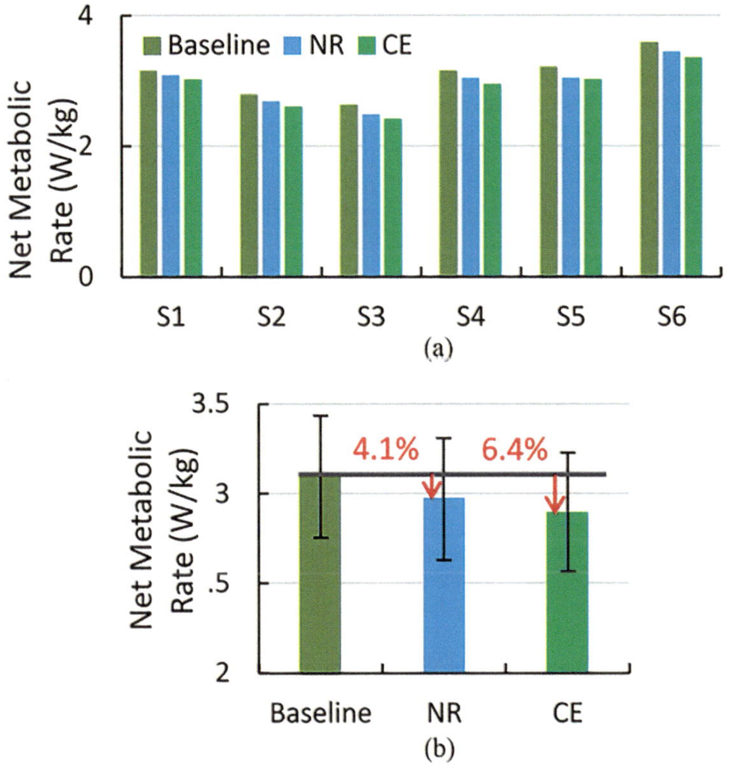

Fig. 10.31 Net metabolic rate of **a** each subject and **b** average of all subjects ($N = 6$) (Reprinted from [19] with permission from IEEE)

by an assistance spring that operates in parallel with the users calf muscles. The experiments demonstrated that the heel-strike energy-storage mechanism enhanced the assistance of the exoskeleton by reducing more metabolic cost compared with walking without the mechanism. We expect this quasi-passive device to highlight a promising basis for the future development of exoskeleton technology.

References

1. Makinson B (1971) Research and development prototype for machine augmentation of human strength and endurance. Hardiman I Project. General electric company report,
2. Yan T, Cempini M, Oddo C et al (2015) Review of assistive strategies in powered lower-limb orthoses and exoskeletons. Robot Auton Syst 64:120–C136
3. Chu A, Kazerooni H, Zoss A (2005) On the biomimetic design of the Berkeley lower extremity exoskeleton (BLEEX). In: Proceedings of the 2005 IEEE international conference on robotics and automation, ICRA 2005. IEEE

4. Guizzo E, Goldstein H (2005) The rise of the body bots [robotic exoskeletons]. IEEE Spect 42(10):50–56
5. Peshkin M, Brown D, Santosmunne J et al (2005) KineAssist: a robotic overground gait and balance training device. In: International conference on rehabilitation robotics, pp 241–246
6. Asbeck A, De Rossi S, Holt K et al (2015) A biologically inspired soft exosuit for walking assistance. Int J Robot Res 34(6):744–762
7. Lee S, Karavas N, Quinlivan B et al (2018) Autonomous multi-joint soft exosuit for assistance with walking overground. In: 2018 IEEE international conference on robotics and automation (ICRA). IEEE
8. Nuckols R, Lee S, Swaminathan K et al (2021) Individualization of exosuit assistance based on measured muscle dynamics during versatile walking. Sci Robot 6(60):eabj1362
9. Hidayah R, Jin X, Chamarthy S et al (2018) Comparing the performance of a cable-driven active leg exoskeleton (C-ALEX) over-ground and on a treadmill. In: 2018 7th IEEE international conference on biomedical robotics and biomechatronics (Biorob). IEEE
10. Park Y, Chen B, Young D et al (2011) Bio-inspired active soft orthotic device for ankle foot pathologies. Intell Robots Syst 4488–4495
11. Collins S, Wiggin M, Sawicki G (2015) Reducing the energy cost of human walking using an unpowered exoskeleton. Nature 522(7555):212–215
12. Ishikawa M (2005) Muscle-tendon interaction and elastic energy usage in human walking. J Appl Physiol 99(2):603
13. Wiggin M (2014) Characterizing optimal performance of a passive elastic ankle exoskeleton during human locomotion. North Carolina State University
14. Gan Z (2018) Design of a lower limb rehabilitation exoskeleton drived by tendon-sheath artificial muscle. Southeast University
15. Collins S, Wiggin M, Sawicki G (2015) Reducing the energy cost of human walking using an unpowered exoskeleton. Nature 522(7555):212–215
16. Brockway J (1987) Derivation of formulae used to calculate energy expenditure in man. Hum Nutr Clin Nutr 41(6):463–471
17. Wu Q (2016) Research on gravity balanced upper limb rehabilitation exoskeleton using tendon-sheath actuator. Southeast University
18. Wang C (2022) Research on lower-limb exoskeleton based on passive dynamic walking theory. Southeast University
19. Wang C, Dai L, Shen D et al (2021) Design of an ankle exoskeleton that recycles energy to assist propulsion during human walking. IEEE Trans Biomed Eng 69(3):1212–1224
20. Zhang Q (2021) Research on design and movement control of tendon-sheath artificial muscles for exoskeleton applications. Southeast University

Chapter 11
Current Issues and Future Perspective

11.1 Current Issues

As a cable-driven transmission method, the tendon-sheath transmission has found wide application in the field of robotics due to its advantages of lightweight, flexible structure, and cost-effectiveness. Compared to traditional high-load drive mechanisms such as electric motors with gears and hydraulics, tendon-sheath transmission offers substantial load-bearing capability with lightweight construction and high flexibility, such that they are popular in exoskeleton robots and flexible robot arms. In applications requiring high precision, speed, and torque, such as industrial automation and collaborative robots, tendon-sheath transmission allows for extensive motion range, and flexible power transfer, making it suitable for joint control in robot arms. Additionally, tendon-sheath systems are well-suited for constructing compact, lightweight, multi-degree-of-freedom motion mechanisms, such that they are widely applied in slender robots. The slender robots typically need to perform precise operations in confined spaces, and the compact multi-degree-of-freedom mechanisms constructed with tendon-sheath transmission are suitable for executing accurate tasks in such applications, as seen in endoscopy, where tendon-sheath transmission can be used for precise treatments and sampling. In Chap. 9 of this book, we provide a detailed introduction to the tendon-sheath structural design for multi-degree-of-freedom slender robots and medical endoscope robots.

It can be said that lightweight and flexible transmission are the significant advantages of tendon-sheath transmission systems, which are also the main reasons for their widespread application in various robotic systems. Among the robots involved, we believe the following directions are currently worth paying attention to.

X. Wang and Q. Zhang, *Tendon-Sheath Transmitted Light and Slender Robots*,
https://doi.org/10.1007/978-981-96-5581-6_11

11.1.1 Miniaturization for Intervention

Surgical robotic systems employing tendon-sheath mechanisms enable precise and minimally invasive surgical procedures. This contributes to reducing trauma and bleeding during surgery while also shortening the recovery period. Nevertheless, the pursuit of even smaller surgical incisions and finer pathways within the human body has been an ongoing objective in the field of surgical robotics. The ability to adapt to narrow and curved entry pathways while maintaining adequate surgical dexterity represents an urgent demand for expanding the scope of intracavity surgical robot applications. In the design of such robots, the transmission, sensing and control technologies in such narrow and complex channels are the keys in meeting the application requirements.

Tendon-sheath mechanisms inherently possess the advantages of small size and flexibility. However, in the applications of the intracavitary robots with multi-degree-of-freedom end-effectors, these systems often incorporate multiple parallel arrangements of tendon-sheath mechanisms. This can result in several undesirable consequences, such as an increase in the overall radial dimensions of the robot, a reduction in flexibility, an increase in weight, diminished adaptability to natural cavities, heightened patient discomfort, and decreased surgical safety. Furthermore, the slender tendon-sheath systems can lead to lower rigidity in the robot. While reduced rigidity is advantageous for the robot to navigate through curved natural cavities, excessively low rigidity can limit the stability of the robot during end-effector operations or in certain scenarios. The flexible design of tendon-sheath systems may also introduce vibrations during operation, potentially impacting operational precision.

Therefore, a promising approach to downsizing surgical robots is the utilization of multi-tendon-sheath transmission methods with finer dimensions, enhanced flexibility, and adjustable stiffness. To achieve this, we need to conduct research involve the selection of lightweight materials, improvements in tendon-sheath shapes and dimensions, optimization of multi-channel sheath structures to reduce the overall radial dimensions and weight, and enhancement flexibility, and achieve adjustable stiffness designs. This presents an effective path for further miniaturization and broadening the application scope of single-port and natural orifice surgical robots.

Furthermore£ the modeling of the transmission characteristics of multi- tendon-sheath mechanisms, the transmission paths sensing, and the utilization of advanced control methods are also the focus of subsequent research. On this basis, the multi-tendon-sheath transmission technology can meet the pressing demands for minimally invasive surgical robots in fields such as neurosurgery, head and neck surgery, pediatrics, and other confined space applications.

11.1.2 Lightweight and High Power Transmission for Exoskeleton

As the demand for medical rehabilitation and elderly care increases globally, exoskeletons and prosthetics have emerged as a crucial technological solution to address the global healthcare needs. To date, various prototypes of exoskeletons have been developed. However, existing rehabilitation and assistive robot devices have challenges such as large physical dimensions, inadequate flexibility, limited portability, and insufficient battery life to meet the daily mobility requirements. Therefore, the lightweight, compact, flexible, and highly capable exoskeletons that offer comfortable wear has become the development objective. In this context, the drive and transmission mechanisms of exoskeletons have become a key focus for overcoming these challenges. Scholars from various countries have contributed significantly to this field, with a wide range of original research, including developments in pneumatic muscles and material-based artificial muscles, being widely applied in various robotic systems. However, as the demand for coordination, stability, and human-machine collaboration in exoskeletons and prosthetics, artificial muscles show limitations such as nonlinearity, low contraction ratios, poor biomimicry, and inadequate multi-joint coordinated driving capabilities, presenting significant challenges in the development of muscle-like actuators.

Tendon-sheath mechanisms offer the ability to accommodate complex curved paths, making them suitable for constructing exoskeletons that can closely fit the human body. This enables users to walk and move more naturally, and tendon-sheath systems are relatively lightweight, contributing to a reduction in the weight of exoskeleton devices. However, when developing lightweight and high-strength exoskeletons using tendon-sheath mechanisms, several challenges need to be addressed. Firstly, one of the drawbacks of tendon-sheath systems is that they introduce some delay in force and displacement transmission. This delay may affect the precision and stability of motion control in exoskeletons and is a common issue of tendon-sheath mechanisms, which requires precise models and advanced control algorithms. Additionally, the arrangement of external sheaths may limit the range of motion in human joints, leading to discomfort or even joint injuries. Therefore, in the design of tendon-sheath exoskeletons, it is essential to consider the characteristics of human motion, design joint forms and cable transmission paths sensibly to maximize the fulfillment of joint motion requirements.

Moreover, tendon-sheath mechanisms typically rely on electric motor actuation, which results in a substantial need for motors and power supply in multi-joint exoskeleton devices, leading to high driving costs, insufficient battery life, and increased overall device weight. Addressing this challenge, the incorporation of combined active and passive degrees of freedom presents an effective means to reduce mechanism complexity and lower driving costs. For example, some lower limb exoskeletons consider human kinematics by setting the ankle or knee joints as passive joints to reduce the usage of driving motors. Alternatively, they may harness the characteristics of human walking by employing a single motor to drive the coordinated movement of two leg joints.

Nonetheless, the human skeletal-muscular system constitutes a complex motion system, and human movement is a result of multi-joint coordination, primarily reflecting the coordinated contractions of various muscles. The introduction of passive joints, while simplifying the drive mechanism and reducing driving costs, can lead to uncertain joint positions, and the absence of active degrees of freedom may not fulfill the multifunctional motion requirements of complex joints. To address this, we propose a solution involves using a single motor and multiple clutches to achieve multi-joint drive control, such that to establish a multi-joint drive system based on the principles of human multi-joint coordination. The multi-cable-clutch transmission mechanism offers a more flexible drive structure, lower driving costs, and reduced dependence on multiple motor devices, effectively meeting the multifunctional motion demands of exoskeletons. This advancement contributes to the progression of exoskeletons and prosthetics towards improved light weight, low cost, comfort, and functionality.

11.1.3 Modeling and Control of Tendon-Sheath Transmission

Modeling of tendon-sheath systems serves as a fundamental aspect of their research and application. Researchers employ mathematical modeling to describe the nonlinear characteristics of tendon-sheath systems, such as cable bending, stretching, relaxation, and transmission friction. These models are instrumental in predicting the force and displacement output of tendon-sheath systems under various conditions, including different speeds, transmission paths, and friction scenarios. Hence, modeling the transmission characteristics plays a vital role in the optimization of cable-driven system designs and motion control.

In this book, we introduce static transmission characteristic models and dynamic transmission characteristic models for various tendon-sheath systems. On this basis, we introduce model-based tendon-sheath nonlinear control methods and advanced control algorithms to achieve high-precision control of tendon-sheath systems. These methods are designed to correct nonlinear effects, thereby enhancing motion accuracy and stability. Furthermore, this book provides detailed discussions on two specific measurement methods as cable bending angle sensors: one based on changes in cable bending displacement and the other based on Fiber Bragg Grating (FBG). Developing relevant sensors and control methods is an important development direction to improving real-time control accuracy.

11.2 Future Perspective

In tendon-sheath transmission applications, in addition to taking the inherent transmission advantages, we believe that enhancing the functionality and intelligence of tendon-sheath systems is the key to expanding their application in newer and more powerful robotic systems. Currently, there is a growing focus on research in robot actuation, characterized by higher power density and integrated sensing systems, often referred to as artificial muscles. Researchers are exploring smart materials and flexible structure designs for their potential in artificial muscles and bio-inspired actuation, including shape memory alloys, carbon nanofibers, electroactive polymers (EAPs), liquid crystal elastomers (LCEs), and pneumatic/hydraulic muscles. The properties of these materials, such as deformability, controllability, and response speed, make them promising choices for developing advanced integrated drive systems. These artificial muscles can not only achieve the motion transmission of traditional actuators but their flexible structures, variable configurations, and high power density ratios make them more suitable for various micro, lightweight, bio-inspired, high-speed robotic systems, and they also open up new avenues in the field of bio-inspired robotics. Although tendon-sheath mechanisms have already demonstrated their advantages in robotic applications, there is still ample potential to explore in terms of increased adaptability, enhanced intelligent functionality, and more biomimetic cable-driven systems. We believe that tendon-sheath system can enhance their driving capabilities in the following aspects in the future.

11.2.1 Artificial Intelligence in Modeling and Control of Tendon-Sheath Systems

Artificial intelligence (AI) methods represented by machine learning have shown significant potential in the modeling and real-time control of nonlinear transmission systems. These advanced technologies can address the complexities and dynamic behaviors inherent in such systems, offering several promising avenues for future development.

As the tendon-sheath transmission is highly nonlinear and its transmission characteristics change in real time with the curved casing path, researchers have tried to predict the transmission output using learning algorithms such as neural networks, and some researchers have used image recognition to identify the bending path of the sheath to compensate for the dynamic transmission loss. In addition, the use of data-driven control methods to improve the motion control of the tendon-sheath-driven robot system has also been reported. These method can leverage vast amounts of data to create highly accurate models. Unlike traditional models that rely on predefined equations, data-driven models can capture the intricate relationships and dependencies in the system. What's more, machine learning algorithms can continuously adapt and update models based on new data, ensuring that the models remain accurate over time as the system evolves or as operating conditions change. Furthermore, these methods can implement predictive control strategies that anticipate future states of the system and make real-time adjustments. This can significantly improve the performance and efficiency of nonlinear transmission systems, and are robust to uncertainties and disturbances, providing stable and reliable control under varying conditions.

However, the parameter identification and optimization is time-consuming and computationally intensive, therefore, in order to better cope with real-time systems, faster and lighter learning algorithms need to be developed. In the future research, the control of tendon-sheath actuated systems can combining AI with traditional control methods, such that leverage the strengths of both approaches, resulting in more effective and reliable control strategies. Research into advanced learning techniques such as reinforcement learning, transfer learning, and unsupervised learning will enhance the capability of AI to handle complex and unpredictable system behaviors. In the other way, advancements in hardware and computational capabilities will enable the implementation of complex AI algorithms in real-time, providing immediate responses and adjustments.

11.2.2 Bionic Actuation Configuration of Tendon-Sheath Mechanism

Bionic actuation aims to mimic the movements and behaviors of human and animals and apply them to robotic systems. Tendon-sheath transmission in this field typically involves emulating the biological muscular and tendinous structures to achieve more natural, flexible, and efficient robotic motion. By this means, robots can perform more biomimetic actions, this makes tendon-sheath transmission particularly effective in rehabilitation and assistive devices. To further enhance the naturalness of tendon-sheath-driven robots and the comfort of robot-human interaction, we should start from biological research, gaining a deeper understanding of biological motion and physiological characteristics, then simulating the biomechanical properties and drive configurations of biological muscle, and integrating them with tendon-sheath transmission mechanism.

In the human musculoskeletal system, multi-headed muscles act on multi-degree-of-freedom compound joints such as the shoulder and hip, playing a crucial role in the coordination and functionality of human joint movements. For example, the biceps brachii muscle stabilizes the shoulder joint during elbow flexion and is also a significant muscle involved in the coordinated flexion of the shoulder joint. It is evident that the distribution and contraction state of muscles on the skeleton directly impact the effectiveness of multi-joint coordinated movements. Therefore, to address the above-mentioned issues, based on the characteristics of multi-headed muscle distribution and coordinated contraction, and combined with the advantages of flexible and high-force cable-driven systems, we propose a multi-headed tendon-sheath artificial muscle to realize the multi-degree-of-freedom coordinated actuation. The multi-headed tendon-sheath artificial muscles has multi-branch tendon-sheath mechanism, series elastic elements, and small clutches, it owns multi-branch drive capability. Based on the multi-branch coordinated contraction mechanism of muscles, we can use the multi-headed tendon-sheath artificial muscles to simulate biological muscles in terms of structural features and contraction modes. Such that we can achieve coordinated actuation of multiple degrees of freedom of joints.

Tendon-sheath bionic actuation contributes to a more natural and flexible movements of robots, making them more suitable for performing complex actions in different working environments. and enhancing human-robot collaboration. This results in improved coordination, comfort, and functionality in human-machine interaction systems like exoskeletons and prosthetics. It also increases the motion flexibility, adaptability, and biomimetic characteristics of bio-inspired robots.

11.2.3 Drive-Transmission Integration of Tendon-Sheath System

"Drive-Transmission Integration" is a concept that involves integrating drive systems and transmission systems, aiming to combine power sources with transmission devices to achieve more compact, efficient, precise, and controllable power transmission. This integrated design can be applied in various robotic and automation systems to decrease the size and enhance performance and efficiency. For instance, integrating the motor, reducer, and transmission device into a single unit reduces the size and weight of the system can also be considered as a kind of drive-transmission integration. Such actuation system is particularly advantageous for applications with limited space and lightweight design.

With the development of smart material and artificial muscles, these new types of actuators have achieved high integration in drive-transmission systems that traditional motors cannot attain. It's worth noting that cables are, in essence, a transmission mechanism and currently rely on motors as their driving devices. To further leverage the advantages of tendon-sheath transmission in terms of flexible high-load and long-distance transmission and to highly integrate lightweight driving devices, we can utilize the excellent driving characteristics of smart materials and employ tendon-sheath mechanism as a favorable means of motion transmission, such that developing highly integrated design of tendon-sheath and smart material drivers to achieve high energy density drive-transmission integration.

11.2.4 Drive-Sensing Integration of Tendon-Sheath System

"Drive-Sensing Integration" refers to the integration of drive systems and sensing systems to achieve more precise and intelligent control. This integrated design helps drive systems better perceive their surrounding environment and operational status, allowing for corresponding adjustments and control. In traditional drive-sensing integrated drives, various types of sensors, such as optical sensors, pressure sensors, accelerometers, and temperature sensors are integrated into the drive system to monitor parameters like force and displacement involved in the drive system. Now, innovative drive-sensing integration is more advanced. The drive materials themselves can generate changes in physical quantities like voltage and current due to changes in stress and strain. This drive-sensing integration characteristic has been verified and applied in smart material-based artificial muscles.

To achieve the integration of tendon-sheath sensing and actuation, innovation can be applied to the materials and manufacture used for both the cable and sheath. One promising direction is the development of smart coating sensors. Smart coating sensors are applied to structural surfaces to monitor cracks in the structure. They consist of three layers: a driving layer, a sensing layer, and a protective layer. The driving layer is made of a non-conductive material, and when cracks propagate in the underlying structure, cracks also appear in the driving layer, which in turn causes the sensing layer to crack. The sensing layer is typically composed of conductive materials such as copper or silver, and as cracks expand, the resistance of the sensing layer changes, such that allowing for crack monitoring. These sensors are sensitive to minor structural damage and can be used for detecting the initiation and propagation of cracks in critical aircraft structures, such as fuel tanks, wing structures, and landing gear components. Additionally, the coating sensors can also be developed as electronic skins affixed to robots, prosthetics, or medical devices to simulate the tactile sensing capabilities of human skin.

By combining cable-driven transmission with smart coating sensors, placing them on the cable surface or at specific locations to detect surface strain, sense stress, and bending configurations, we can eliminate the need for tension and displacement sensors, and achieving high integration of cable actuation and sensing. This is particularly crucial for smaller-sized medical robots operating in confined spaces, enabling smarter and more flexible robotic systems.

We think another innovative approach can be explored using FBG (Fiber Bragg Grating) optical fiber sensors. FBG is a sensor based on the principles of fiber optics, commonly used for measuring parameters such as temperature, strain, pressure, and vibration. It features a compact structure and can be easily embedded into the structures being tested, making it widely applicable in engineering, medical, and various other fields. This book also provides a detailed introduction to FBG optical fiber sensors, which can be embedded in long and slender robots to measure cable tension and bending pathways. However, the FBG sensors used in tendon-sheath mechanisms still need a room for installation, increasing the dimensions of the cable-driven structure and decreasing the tendon-sheath flexibility. To achieve sensing and actuation integration in the tendon-sheath transmission, we can upgrade the tendon-sheath mechanism by developing a hybrid tendon-sheath grating system. By inscribing gratings (both optical and magnetic) on the cable, it becomes possible to sense the stress and strain of the inner cable. By strategically planning measurement locations, bending configurations of the tendon-sheath can also be sensed.

11.2.5 More Tendon-Sheath Applications

Tendon-sheath transmission mechanism offers precise motion control, adaptability, and reliability for robots to perform a variety of tasks. In the future, tendon-sheath transmission mechanism can be used in more application areas:

(1) In the industrial robotics applications, tendon-sheath transmission that has relatively lightweight can reduce the overall weight of the robot, enhancing its maneuverability and speed. Additionally, it enables precise joint position control, making it useful for tasks such as automated assembly, welding, and part handling.
(2) Tendon-sheath are also suitable for underwater robot applications. The slender and flexible structure of tendon-sheath transmission allows underwater robots to adapt to different working environments, such as inspecting underwater pipelines or conducting underwater scientific research.
(3) In the aerospace field, tendon-sheath transmission can be utilized for space robots and satellite to carry out space exploration, sampling, satellite orientation adjustments. The low-load characteristic of tendon-sheath systems is vital for space applications as it can reduce the payload and energy requirements of the satellite.
(4) In military applications, the long-distance transmission capability of tendon-sheath mechanism can be used for military robots to achieve remote operations, such as explosive ordnance disposal. Furthermore, combining tendon-sheath transmission with track or wheeled mechanisms can realize more flexible robot configurations for various terrains and performing reconnaissance and inspection operations.
(5) Tendon-sheath can also serve educational and entertainment robots as it has a simple structure and diverse transmission forms, making it easier to understand and teach compared to other transmission systems. Moreover, tendon-sheath mechanism is relatively cost-effective, making it suitable for educational and DIY projects.

However, the tendon-sheath transmission systems that incorporate biomimicry and drive-sensing integration hold the potential to serve more intelligent, functional, and diverse robotic systems. In the future, more efforts will be dedicated to the research and development of intelligent tendon-sheath transmission, and with the attention and collaboration of various fields, including materials science, optoelectronics, signal processing, and robotics, we believe that tendon-sheath transmission will experience a higher level of innovation and a broader scope of development.

Index

X. Wang and Q. Zhang, *Tendon-Sheath Transmitted Light and Slender Robots*,
https://doi.org/10.1007/978-981-96-5581-6

GPSR Compliance

The European Union's (EU) General Product Safety Regulation (GPSR) is a set of rules that requires consumer products to be safe and our obligations to ensure this.

If you have any concerns about our products, you can contact us on ProductSafety@springernature.com

In case Publisher is established outside the EU, the EU authorized representative is:

Springer Nature Customer Service Center GmbH
Europaplatz 3
69115 Heidelberg, Germany

Batch number: 10704307

Printed by Printforce, the Netherlands